FLOATING GOLD

FLOATING
GOLD

A NATURAL (AND UNNATURAL)
HISTORY OF AMBERGRIS

CHRISTOPHER KEMP

The University of Chicago Press   Chicago & London

CHRISTOPHER KEMP is a molecular biologist. He
currently lives in Grand Rapids, Michigan, with his wife
and two sons. He can be contacted at ambergrishotline
@gmail.com.

The University of Chicago Press, Chicago 60637
The University of Chicago Press, Ltd., London
© 2012 by The University of Chicago
All rights reserved. Published 2012.
Printed in the United States of America

21 20 19 18 17 16 15 14 13 12   1 2 3 4 5

ISBN-13: 978-0-226-43036-2 (cloth)
ISBN-10: 0-226-43036-7 (cloth)

Library of Congress Cataloging-in-Publication Data

Kemp, Christopher.
  Floating gold: a natural (and unnatural) history of
  ambergris / Christopher Kemp.
      pages. cm.
  ISBN 978-0-226-43036-2 (cloth: alkaline paper)—
  ISBN 0-226-43036-7 (cloth: alkaline paper)
  1. Ambergris. I. Title.
  QD331.K457 2012
  333.95'95—dc23   2011050637

♾ This paper meets the requirements of ANSI/NISO
z39.48-1992 (Permanence of Paper).

For
EMELINE, MAX, & IZZY,
my fellow ambergris hunters.
This is a record of the time.

In Memory of
DR. ROBERT CLARKE
(1919–2011)

Who would think, then, that such fine ladies and gentlemen should regale themselves with an essence found in the inglorious bowels of a sick whale! Yet so it is. ⋆ HERMAN MELVILLE, *Moby-Dick* (1851)

An ignorant Fellow in *Jamaica*, about two Years ago, found 150 Pound Weight of *Ambergreece* dash'd on the shoar, at a Place in these parts called *Ambergreece Point*, where the *Spaniards* come usually once a Year to look for it. This vast Quantity was divided into two Parts; supposed by Rolling and Tumbling in the Sea. This Man tells me that 'tis produced from a Creature, as Honey or Silk. And I saw in sundry places of this Body, the Beaks, Wings, and part of the Body of the Creature, which I preserved some Time by me. He adds, That he has seen the Creatures alive, and believes they swarm as Bees, on the Sea shore, or in the Sea. ⋆ *Philosophical Transactions and Collections, to the End of the Year, 1700*

Dear Dr. Clarke,

I did have another question for you. I've always pronounced ambergris without the "s" at the end, like the French from which the name is derived. Like this: *ambergree*. Others I've spoken with have pronounced the "s," but softly, like this: *ambergrizz*. How do you pronounce it?

All the very best!
ck

Dear Mr. Kemp,

Like all those with whom I have spoken about ambergris I pronounce the word like this: *ambergreez*, and I recommend that you do the same. After all, you are not French, are you?

Yours,
Robert

CONTENTS

PROLOGUE: WELLINGTON, 2008

On Saturday, September 20, 2008, the excitement was beginning to grow on Breaker Bay, near Wellington, New Zealand. Although it was early spring and still cool, a crowd had gathered to investigate a strange object that had washed ashore during the night. It was large, perhaps the size of a 44-gallon drum, and weighed an estimated thousand pounds or more. No one had seen it arrive. It was just there on the sand. Roughly cylindrical in shape, it was the color of dirty week-old snow.

On the following Monday, national news media were beginning to report the arrival of the object on the beach. To the people who live in the Wellington suburbs near Breaker Bay, the news reports were hopelessly late in coming. Everyone already knew about the object. It had been sitting incongruously on the beach for the last two days. Hundreds of people had already wandered over to take a look at it. Seagulls had been pecking at it as it slowly settled into the sand. Rumors spread quickly through the coastal communities around the bay. It was, one of the prevailing opinions stated, a large piece of cheese. In fact, people said, it was probably Brie. It must have been dumped on the beach by the notoriously rough waters of Cook Strait—a large cylinder of soft cheese swept toward the bay from the busy shipping lanes. Perhaps, some people suggested, it was industrial soap. Others asked: Could it be some kind of meteorite?

And then the situation became stranger still. People were now taking pieces of the object, carving off large heavy servings of it with whatever tools they could find. They then took their samples home, protecting them from decomposition and from the further attention of curious seagulls.

At night I watched the news reports with growing amazement. What would make people act this way? Six months earlier, my wife and I had moved from the United States to Dunedin, a midsize coastal city that sits near the southernmost tip of New Zealand's remote and rain-swept South Island. I was working as a biologist at the University of Otago, a large research institution in the city, and so was my wife. As relative strangers, we wondered aloud to each other if this behavior was particular to New Zealand. Perhaps it's an aspect of the national character, we said, to act this way when something unusual washes ashore.

But another rumor had begun to circulate: The mysterious object was ambergris.

"I'm pretty sure this is ambergris," Geraldine Malloy told a television news reporter breathlessly. "I think it's the find of a lifetime. I've been looking for it for a long time." She was standing over the object, wearing a windbreaker and using a long-handled shovel to break off smaller pieces from the whole.

Before September 2008, I had never even heard of ambergris before. Now I sat transfixed, as people almost rioted on the beach, armed with garden tools, trying to obtain just a small piece of it. It was one of the strangest sights I had ever witnessed, and I was determined to learn more. I discovered that ambergris is excreted only by spcrm whales, and it is rare and extremely valuable. In fact, ambergris is valuable enough that rushing to Breaker Bay armed with a long-handled shovel to break off some smaller pieces from the whole was a reasonable investment of Malloy's time. Used as an ingredient in the manufacture of perfume—and for more esoteric purposes in more distant places— ambergris is traded on the open market for up to $20 per gram, depending on its quality. As a useful reference, gold is currently trading at around $30 per gram. Unlike gold, people occasionally report finding ambergris on remote beaches in lumps that weigh more than fifty pounds—or $500,000 worth. And at $20 per gram, the object being enthusiastically divided up by the crowd at Breaker Bay was worth around $10 million.

Ambergris becomes more valuable as it ages. A well-aged piece of ambergris is unusual, and the arrival of a well-aged piece the size of the strange object on Breaker Bay would have been a singular event. It would have been comparable to finding a lump of gold the size of a suitcase in the middle of a well-traveled path. No one on the beach seemed certain that the object was ambergris, but no one was taking any chances that it wasn't, either. The crowd grew larger. People arrived with bedsheets, which they used as makeshift slings to transport bowling ball-size pieces of the object home. Those managing to secure even a handful—a greasy wet piece weighing maybe a pound or so—were potentially taking home $5,000 worth of ambergris.

The excitement continued to grow throughout the day. "So many people have rung in saying, 'It's worth half a million dollars,'" Wellington City Council spokesman Richard MacLean told the *New Zealand Herald*, "we feel honor-bound to actually go out and stake our claim on it."

And then on Tuesday morning—just three days after its arrival—no sign remained to prove that the object was ever there. The crowd had removed it, piece by piece and pound by pound, with their shovels, bags, and improvised slings. All around Breaker Bay and farther afield, in suburbs across the city, people were celebrating their sudden and unanticipated wealth.

There was only one problem. It wasn't ambergris, says Nic Conland, the Environmental Protection team leader for the Greater Wellington Regional Council. It was a large seaborne block of tallow. "Essentially, it's lard," Conland told me a few months later by telephone. "We're not sure, but it looked like it had been part of a drum that had fallen off a ship."

Pieces of it had started to appear for auction on online trading sites, listed as ambergris. But it was worthless. It was worse than worthless—it was a potentially harmful pollutant. It could clog every kitchen sink in the Greater Wellington metropolitan area. And its provenance was, for the most part, still unknown. On Wednesday—four days after the object had washed ashore—the Greater Wellington Regional Council finally broke its silence, posting an update for the public on its website. It read: "Wellingtonians who earlier this week removed several hundred kilograms of lard from the beach at Breaker Bay are urged to make sure that they dispose of it thoughtfully once they realize that it is not ambergris and therefore largely worthless."

Lard. Largely worthless. The story, for so many, had ended.

For me, it had only just begun.

INTRODUCTION: MARGINALIA

In the course of writing this book, I was asked one simple question again and again: Why ambergris? What was it about the substance, which Franz Xavier Schwediawer had called, in 1783, "preternaturally hardened whale dung," that made me want to trudge month after month along lonely windswept coastlines? Sometimes, especially on wet blustery days, even I struggled to find an answer.

In the beginning, the value of ambergris was as good a reason to go looking for it as any. I was drawn to the idea that I might find something worth $50,000 on the beach, just dumped there, glistening on the high-tide line. For months, this was enough to send me out to the coast. The fact that ambergris looks like an unremarkable piece of driftwood and requires decades at sea to transform just added to its appeal.

But time passed, and my motivations evolved.

When I was a child, I was fascinated by the natural world. My parents fed my interest, buying me books filled with photos of animals. I took them on family vacations with me, so that I could study them during long car journeys, tracing a finger over the giant squid and the Amazonian anacondas. But the most worn and dog-eared pages were always those with the photos of the oddest and most otherworldly animals on them: deep-sea angler fish, duck-billed platypuses, and bird-eating spiders. The marginalia. More than anything else, I was attracted to their strangeness. Years later, on long walks in the English countryside, I collected skulls and brought them home, cleaning them in my bedroom and lining them up alongside one another like trophies: rabbits and squirrels, with their curving yellow incisors; a crow skull, with its sturdy black beak and a brain case as thin and fragile as a Ping-Pong ball.

I can still remember finding a fragment of bone buried deep in the soil in my backyard when I was twelve years old. I sent it to scientists at the nearby Birmingham Museum and Art Gallery with an accompanying letter, asking them to identify it for me. And they did. It was the parietal bone of a fox skull. The reply came with an invitation to visit the museum to see how its scientists prepared specimens for display.

All day my mother and I walked from room to cluttered room, into parts of the museum that were not open to the public. We had passed beyond an invisible barrier, permitted entry to a special exclusive realm. We saw shelves lined with dusty fossil fragments and glass jars filled with enormous preserved insect specimens. At one point I clambered up a stepladder to peer into a bubbling vat of chemicals. Inside, several fox carcasses were undergoing a process to strip the flesh from their bones. As I watched, a technician leaned over the side and used a tool to probe the liquid, hooking a carcass and hauling it to the surface for me to see. My mother—who did not need to see decomposing fox carcasses as badly as I did—nevertheless spent several minutes gently convincing the technicians that it was appropriate to show them to me. And so I stood at the top of the ladder, breathing the warm fumes, looking at the white glistening bones.

It wasn't enough just to visit the museum. I needed to tiptoe through the unvisited parts as well, through the unexamined marginalia not on display. It was a glimpse into a world *behind* a world.

A decade later, I earned a degree in biology; later still, a master's degree in epidemiology. And I embarked on a career in molecular biology and neuroscience. My fascination with the natural world has never left me.

*   *   *

As a scientist, I'm used to being able to access information when I want it. From my desk, I can download millions of scientific articles with the click of a mouse. Within a minute or so, I can know the exact three-dimensional structure of a protein I am studying, or the genetic code that made it.

When I first heard about the mysterious object on Breaker Bay in September 2008, I went online immediately, thinking I'd learn everything I needed to know about ambergris in a few minutes. But I failed. In fact, to begin with I found almost no useful information at all—just a handful of esoteric scientific papers and medical textbooks, most of them published in the eighteenth century. They were full of contradictions and inconsistencies. There are, of course, more recent news articles that mention ambergris, which tend to appear after someone has stumbled over a lump of ambergris on a remote beach somewhere. At least half of these news stories refer to ambergris as whale vomit—a persistent misnomer that suggests journalists are not taking their work seriously. They have headlines like "WHALE COUGHS UP A JACKPOT"—an article from the *New Zealand Herald* in 2006. A few of the more complete whaling histories include brief sections on ambergris—notably *The Natural History of the Sperm Whale* by Thomas Beale (1839), *Whales and Modern Whaling* by James Travis Jenkins (1932), and most recently *Leviathan: The History of Whaling in America*, by Eric Jay Dolin (2007). But mostly, it is marginalia.

It wasn't always the case. In 1794, when Henry Barham wrote *Hortus Americanus*, exclaiming that ambergris was "a universal cordial," it was one of the most widely used substances in the world. But that was 1794. Apart from a few traders and perfumers and a handful of fortunate beachcombers, the world seems to have forgotten about ambergris. For a few weeks toward the end of 2008, I had almost forgotten about it too. I'd managed to put it out of my mind. My wife had just given birth to our son. We were sleep-deprived and busy. Whenever we could, though, we drove along the winding harbor road, past clumps of flax and cabbage trees, to walk the shoreline of the local beaches.

I would follow the high-tide line—a long, irregular wet trail—pushing my son's stroller across the beach, collecting objects from the sand. Just like when I was a child, I was less interested in the easily identified objects—the tapered green fronds of bladder kelp and the empty mussel shells. Instead, I was drawn to the strange and unidentifiable, collecting odd and misshapen items that might have come from the other side of the world.

Gradually, I began to think of ambergris again. I started to search for it along the shoreline. And soon enough, I was calling museum curators in London and tracking down international ambergris traders. It was as if I had fallen down a rabbit hole. More than anything else, my motivation was the complete lack of reliable information. I could no longer stand not knowing. The scientist within, the part that searches for clarity, wouldn't let me rest until I had discovered the truth. Even the descriptions of the characteristic odor of ambergris seemed inadequate. Surely, I reasoned, it was simply that no one had tried hard enough yet to describe it. Eventually, it became more than I could bear. I decided to do everything possible to experience and describe ambergris properly. What was the point, after all, of reading about the unmistakable odor of ambergris if I could not then smell it for myself?

As a result, I spent two years exploring some of the strangest marginalia I had ever come across. In the process, once again, I walked through the unvisited and forgotten corners of familiar places, and glimpsed secret worlds. I was driven by the same irresistible impulses I'd had as a twelve-year-old child. I wanted, in other words, to experience something more fully and completely than anyone else. And the need to do so took me across the world, in search of ambergris.

1 ON LONG BEACH

Ambergris is an extraneous Substance, that swims in the Sea, and is swallowed as a Delicacy by the Fishes, and voided by them again undigested. It seldom stays long enough, to be found in their Bodies. * CASPAR NEUMANN, "On Ambergris" (1729)

I tell people that they've got to sniff a lot of dog droppings before they find a bit of ambergris. * Interview with amateur historian LLOYD ESLER, New Zealand (2010)

* * *

It's a rainy afternoon on Long Beach. I am standing beneath a mackerel sky, holding a strange little object in my hand. It's a pale green-gray color, like a barely steeped cup of green tea, and it looks like a potato. I hold it up to the gray light and examine it more closely in the rain. Sitting in the palm of my hand, it feels light and spongy. It could be a thick stalk of decomposing seaweed, still wet from the ocean, or an old and waterlogged piece of driftwood. It might be a shriveled piece of marine sponge, dislodged from the seafloor and then washed ashore by the last tide. It could be an almost infinite number of different things. In fact, the object in my hand could actually *be* a potato. It might have traveled from the other side of the world, bobbing and rolling around on ocean currents for months, or even years, before finally arriving on the beach. I bring it to my nose and carefully smell it, hoping it is ambergris. Nothing. It has no detectable odor, except perhaps the faintest briny trace of the sea. And so I discard it and move on again, slowly making my way northward along the beach. Head lowered, I survey the wet sand, bending occasionally to pick up an object before smelling it and then pitching it over my shoulder. Behind me, I have left a wide and messy debris field.

* * *

Long Beach is a remote 1.5-mile-long strip of sand located almost fifteen miles north of Dunedin, a city that sits near the southern tip of the South Island of New Zealand. It is well named: a long straight shot of sand, like a thin yellow band wedged between the sea and the towering cliffs. At the tidemark, a long skein of bladder kelp: wrist-thick green cables terminating in large disc-shaped holdfasts, dumped unceremoniously on the sand by the last tide. This is ambergris territory. Remote. Windswept. White noisy waves crash along the length of the sands.

I have come to Long Beach to find ambergris, a substance I have never seen or smelled before. Even if I see a piece of it, half hidden by a wet tangle of bladder kelp, I will probably walk past it. Despite this fundamental obstacle, I reassure myself every few paces that I will find some ambergris eventually if I am simply willing to spend long enough searching for it. And I trudge onward through the rain: cliffs unspooling solidly on my left, and the sea, a shifting range of slippery gray peaks, to my right.

Everything I know about ambergris, I have learned from watching the news reports on the enormous drum-shaped lump of lard that washed ashore on Breaker Bay in 2008, which means I know almost nothing. In fact, I know less than nothing, because the reports had been filled with inaccuracies.

But I've learned a few important things: first, ambergris is an intestinal secretion, expelled only by sperm whales. It washes ashore with the tide and has a complex and hard-to-describe smell. I also know that ambergris has been used for centuries to make perfume. It acts as a fixative, anchoring the fragrance to the wearer's skin and making it last longer. Finally, and most importantly, I have read that ambergris is valuable. In fact, it's worth so much that hundreds of people had descended onto Breaker Bay, dismantled a half-ton block of lard with their garden tools, and then taken pieces of it home, believing it was ambergris.

At home, I had begun to research ambergris: I visited libraries, leafed through encyclopedias, took notes from textbooks on marine mammals, and copied recipes from old perfume formularies. I read old ledgers, journals, and court registers. And I spent an inordinate amount of time online, reading dense scientific literature and trying to understand strange 400-year-old texts. I had learned, for instance, that when King Charles II's daughter Elizabeth was born in St. James Palace in 1635, "the states of Holland, as a congratulatory gift to her father, sent ambergris, rare porcelain and choice pictures." And in 1689, when English philosopher John Locke published his landmark *Two Treatises of Government*, he used ambergris to make a point about ownership, writing: "By virtue thereof, what fish any one catches in the ocean, that great and still remaining common of mankind; or what ambergrease any one takes up here, is by the *labour* that removes it out of that common state nature left it in, *made* his *property*, who takes that pains about it." In 1693 the Dutch East India Company bought an enormous lump of ambergris weighing 182 pounds from the king of Tidore—a small and remote Indonesian island kingdom—and the Grand Duke of Tuscany, on the other side of the world, wanted it so badly for himself that he offered fifty thousand crowns for it. A seventeenth-century writer traveling through Persia wrote, "The usual drink is sherbet made of water, juice of lemmons and ambergreece." And Casanova, I read, added ambergris to chocolate mousse, and then ate it for its alleged aphrodisiac qualities. But almost all the references I have found are historical: strange ephemera, weird facts, miscellanea, and curiosities from old books. I have begun to believe that ambergris is something that belongs to the past.

And then I find another, much more recent, account:

On Tuesday, May 9, 2006, ten-year-old Long Beach resident Robbie Anderson was walking his dog Scud on the beach—on *this* beach, not far from where I'm standing, amid a field of already-rejected debris. Making his way along the shoreline, past windblown sedges and shiny clumps of flax, he found a piece of ambergris in the sand. At the time, Anderson didn't even know what he had found: it could be a dirty piece of soap, he told himself, or maybe it was part of a decomposing sheep carcass, washed ashore and beached by a recent storm. It was waxy and smelly and strange, but he decided to keep it anyway. So he retrieved it from the sand, tucked it under his arm like a loaf of fresh bread, and carried it home, where he presented it to his father.

It was unremarkable in appearance: a mottled white and gray color, irregular in shape, slightly flattened, and about the size of a football. But it had a strong odor, which was unusual and difficult to categorize. Half buried by sand near the tide line, it had looked like a rotting tree stump or a charred piece of driftwood. I would have walked straight past it on the beach without giving it a second glance. From a distance, I might have mistaken it for a dead seagull or a waterlogged shoe. But it was ambergris It weighed about a pound and a half. The following day, Robbie and his father—who is also called Robert—returned to the beach and carefully scoured the shoreline. By then, they had researched the object and realized it could be ambergris, worth a lot of money. On the second day, they found an additional half pound of ambergris, broken into several smaller pieces, which they took home and placed next to the larger lump. The total haul was worth approximately $10,000.

A brief report of the discovery was published a few days later in the *New Zealand Herald* under the headline "WHALE COUGHS UP A JACKPOT." Accompanying the article was a photograph of Robbie Anderson on the beach: a wide toothy grin on his face, ambergris cradled in his hand like a chunk of wet rotten wood. Since reading about the find, I have visited Long Beach often. It's why I'm here now, walking along the tide line in search of ambergris and hoping that, by simply wanting it enough, I can somehow will another two-pound lump ashore so that I can find it, among the drifts of kelp and the empty crab shells, overturned in the sand.

* * *

Ambergris begins its long journey in darkness, beneath several hundred tons of seawater, in the warm and cavernous hindgut of a sperm whale. This much is known. If many aspects of its journey are not so clear, this is because the lives of sperm whales are still mostly shrouded in mystery— a collection of theories rather than facts. As they leave the ocean surface, glistening green-gray flukes disappearing beneath the chop, they simply leave our world behind and dive into another.

Measuring up to sixty feet in length, an adult male sperm whale is the largest of the toothed—or odontocete—whale species. Like most whale species—such as the barnacled southern right whale, with its huge arched grill of baleen; or the beluga whale, with its bulbous head—the sperm whale is a strange-looking animal. Its eyes seem hastily and carelessly placed, almost marooned behind the blunt box-like head, located up to a third of the way along its tapering torpedo-shaped body. Or, as Herman Melville wrote in *Moby-Dick*: "Now, from this peculiar sideway position of the whale's eyes, it is plain that he can never see an object which is exactly ahead, no more than he can see one exactly astern. In a word, the position of the whale's eyes corresponds to that of a man's ears; and you may fancy, for yourself, how it would fare with you, did you sideways survey objects through your ears."

To maintain its prodigious body weight—approximately fifty tons for a bull and around twenty tons for a cow—a sperm whale must consume about a ton of food a day, diving again and again to pressures that collapse its flexible rib cage to a quarter of its normal volume. In fact, almost everything about a sperm whale is as implausibly oversize as its appetite: a seventeen-pound brain; a lower jaw punctuated with around fifty large, conical teeth, each of which weighs up to two pounds; a huge muscular heart. In a 1959 article from the scientific journal *Circulation* titled "A Large Whale Heart," researchers described a 256-pound sperm whale heart, which was removed from a large bull whale killed by a Peruvian whaling company. The heart is a large marbled mass, like a misshapen mound of uncooked dough, striated with bands of muscle and dotted with thick valves. One of the authors holds the large vessels of the heart in his hands for a series of grainy black-and-white photographs. He points the open gaping end of the aorta toward the camera. It measures almost eight inches across and resembles an empty pant leg.

During just one of its hour-long dives, beyond the farthest reaches of sunlight to depths of more than a mile below the ocean surface, a large bull sperm whale can ingest hundreds of pounds of deepwater squid. After more than twenty minutes shouldering steadily downward, through progressively colder layers of water to the mesopelagic zone— between 650 and 3,300 feet below sea level—he feeds gluttonously on cephalopods near the seafloor, eating squid that range from just a few ounces in weight to huge muscular specimens weighing more than two hundred pounds.

This much is known, but so much regarding the daily activities of sperm whales—a significant proportion of which occurs at extreme pressure, deep below the ocean surface—is not. It remains a complete mystery. In some respects, we know as much now as we did in 1770, when James Robertson, Esq., of Edinburgh, described a beached dead sperm whale—referring to it as a "cachalot," a name given to it by the French— for the journal *Philosophical Transactions* in an article titled "Description of the Blunt-headed Cachalot":

PHYSETER, *Catodon Linnaei*, blunt-headed Cachalot, British Zoology, run ashore upon Cramond Island, and was there killed, December 22, 1769. Cramond Island is in the Firth of Forth, four miles above Leith The fish measured fifty-four feet in length; its greatest circumference, which was a little behind the eyes, thirty. The head was nearly one half the whole fish, of an oblong form, and rounded, except within six feet of the extremity, where it had inequalities, shewn by the transverse section.

Robertson went on to describe the carcass in detail: its tapered body and wedge-shaped tail; its toothless upper jaw with "twenty-three sockets on each side, for lodging the teeth of the lower, when the mouth was shut"; and its "remarkably small" eyes. He details the spermaceti organ— two large oil-filled reservoirs, crisscrossed and laced with capillaries— which is responsible for the sperm whale's large blunt head and occupies most of its volume. As A. F. Busching noted in 1762, "The head makes near half the bulk of the fish, not unlike the butt end of a musket."

During the whaling era, a large bull whale could yield as much as four tons of valuable oil. Prized by whalers as lamp oil, as a material for candle making, and used worldwide as a commercial lubricant, the oil turns

white and congeals on contact with the air. It earned the whale its name
when it was first mistaken for semen. Robertson wrote: "The substance,
improperly called *Spermaceti*, and erroneously said to be prepared from
the fat of the brain, was everywhere contained in a fluid state in the cavity of the head along with the brain, but quite distinct from it."

Back in 1770, when Robertson was describing the sperm whale, the
purpose of the spermaceti organ was unknown. The value of the oil contained within it was not. "To come at that fluid, the workmen made a hole
into the cavity of the head," Robertson explained, "and took it out with a
skimmer from among the substance of the brain, as it flowed to the hole,
which it did like water springing up into a well." Almost 250 years later,
the purpose of the spermaceti organ is still unknown. Whale experts—or
cetologists—study such things aboard research vessels that bob around
on the ocean surface, miles above their elusive subjects. They might as
well be using their instruments to study the geology of Mars.

Among the remaining mysteries: How does such a lumbering and
slow-moving mammal manage to eat so many squid? Are the squid
sluggish and vulnerable at such extreme depths and temperatures? Does
the pale pigmentation of the whale's lips attract squid toward its open
mouth? Do the whales stun squid momentarily defenseless with a sudden burst of sonar clicks? No one knows. Almost all theories are possible
and worthy of consideration.

*   *   *

Sperm whales supplement their cephalopod-rich diet with benthic crabs
and octopuses, and with rays and other large fish, including sharks measuring up to twelve feet long. Their energy requirements preclude fussi-
ness. They are huge engines, burning fuel constantly. A highly organized
species, they break the underwater silence, communicating with flurries
of clicks and vocalizations. Their communications are sophisticated and
complex, but nothing we can understand. They have been observed co-
ordinating their feeding efforts, with several whales fanning out in a
wide arc half a mile long to hunt—holding their positions more than
a hundred fathoms beneath the surface, rounding up and herding their
prey in the deep-sea gloom. Their diet is a reflection of their surroundings: near Iceland and in the cold waters of the Gulf of Alaska, they eat
fish almost exclusively; and farther south, in the warmer waters of the
Azores, their diet consists mostly of squid.

In a 1993 study, cetologists surveyed the stomach contents of seven-

teen sperm whales killed by commercial whalers in the Azores and reported a total of just sixteen fish among the half-digested remains of
almost twenty-nine thousand squid. In the digestive system, a squid is
broken down quickly, leaving undigested only the mouthparts—called
a beak because of their resemblance to a parrot's beak—along with the
inflexible and indigestible eye lenses, and a tough internal quill-like organ called the pen. In the opened and dissected stomachs sat the durable
beaks of at least forty different species of squid: mostly from the Octopo-
teuthidae, Histioteuthidae, and Architeuthidae families, with a few from
the Lepidoteuthidae and Ommastrephidae families and smaller numbers of numerous other species thrown in for good measure.

In the belly of a whale, a single solitary squid beak can tell a complex
tale. In a general sense, it represents a crude three-dimensional map of
the world, with different species of squid occupying distinct and limited
geographic regions. Even within those well-defined regions, some squid
species are present at some depth zones and completely absent from others. In other words, a lone squid beak, trapped in the warm folds of a
whale's stomach, removed by cetologists, can help provide a history of
movement. In the same way that one can tell where a letter was mailed—
even after the envelope has ceased to exist—by studying the marks imprinted on its faded postage stamp, whale researchers can learn a lot
about a whale from the squid beaks found in its belly. The presence of
beaks from *Megalocranchia* or *Gonatus* squid genera in the stomach of a
sperm whale killed in Azorean waters can mean only one thing: movement. These species do not belong anywhere near the Azores, a Portuguese archipelago in the Atlantic Ocean. They thrive thousands of miles
to the north, in colder waters. And, removed from the dissected stomach
of a sperm whale in the subtropical Atlantic, *Megalocranchia* and *Gonatus*
squid beaks indicate purposeful movement across vast distances.

Squid beaks are important for another reason: they help to produce
ambergris. As Robert Clarke—the preeminent world expert on ambergris—explains in "The Origin of Ambergris" from 2006, included in
the ton of squid a sperm whale eats daily are several thousand squid
beaks. Like cows and other ruminants, a sperm whale has four stomachs. Food passes from one stomach to the next and is digested along the
way. Steadily, after repeated dives and bouts of voracious feeding a mile
beneath the surface, the stomachs slowly begin to fill with nondigested
squid remains: great drifts of sharp, black, durable squid beaks, which co-
alesce to form a large dense glittering mass. Every couple of days, a sperm
whale will vomit them into the ocean. This is normal. Importantly, the

product, a floating slurry of indigestible material, is not ambergris. It is whale vomit. The two could not possibly be confused with one another. Despite newspaper headlines to the contrary, ambergris is not vomited or coughed up by sperm whales. Robbie Anderson's ambergris—picked up on Long Beach, tucked beneath his arm like a warm baguette, and taken home to Robert, Sr.—was not "spat out by a sperm whale that swam past coastal Otago," as stated in the *New Zealand Herald* news report a few days later. To produce ambergris, other processes—complex pathologies—are required. Occasionally, the mass of squid beaks and pens makes its way through each of the whale's four cavernous stomachs and into its looping convoluted intestines instead. Once there, it can become ambergris.

Reading parts of Clarke's "The Origin of Ambergris," I find it impossible not to imagine him sitting across from me in the galley of a pitching and yawing vessel, several days from landfall and dimly lit by a guttering lantern, as we slide down the sheer gray face of another thirty-foot wave somewhere in the southern Pacific. He wrote:

> Now once in the Antarctic in 1948 on board Fl. F. *Southern Harvester* I examined a sperm whale whose cylindrical last stomach was entirely filled with a compacted mass of squid beaks, squid pens and nematode worms. The mass was 1.2 m in length and 0.4 m in diameter. This last stomach is normally empty except for a few small beaks, pens and nematode cuticles. We have only to imagine an imperfect valve, a leaky sphincter between this last stomach and the intestine, when all conditions are set for a train of events which should result in ambergris.

This occurs, Clarke estimated, in around 1 percent of sperm whales. Current sampling methods—which are inexact and always debated—put the sperm whale population at approximately 350,000 worldwide, which means ambergris is produced by only 3,500 sperm whales, scattered throughout the world's oceans. This explains its rarity—its singularity.

Curved like a parrot's beak, the squid beaks pass from the stomach, chafing and irritating the delicate intestinal lining on the way. As a growing mass, they are pushed farther along the intestines and become a tangled indigestible solid, saturated with feces, which begins to obstruct the rectum. It acts as a dam. Feces build up behind it. The whale's gastrointestinal system responds by increasing water absorption from the lower intestines, and gradually the feces saturating the compacted mass of squid beaks become like cement, binding the slurry together permanently. It becomes a concretion—a smooth and striated boulder. Temporarily, feces make their way past it again, passing between the boulder and the wall of the intestines. And, slowly, the process repeats, adding additional strata to the boulder, which grows larger with each new layer in the same way that a tree grows, adding a new growth ring with each passing year.

Perhaps, in some instances, a whale is able to pass the ambergris. In others, the growing boulder of ambergris is fatal. It occludes the gut completely, Clarke explained, and the whale suffers a fatal intestinal rupture. In a process that takes years, one stratum too many has been laid over the top of the others. The ambergris has grown too large for the gut. The dead whale, now adrift on the open seas, slowly begins to swell. Within hours, the stinking carcass will be surrounded by sharks, drawn to the blood in the water like iron filings to a magnet—makos and blues, mostly. From the air, gulls, storm petrels, and shearwaters will arrive in a noisy tangle and settle in the water around the bloated corpse, which has begun to trail a greasy slick of oil behind it through the waves. The smaller fish will feed on it from below, tearing the flesh into strips and fighting over it among themselves. At some point, the ruptured intestines will be torn open by scavengers, and the ambergris will fall into the ocean. The whale carcass will become a floating bounty of food in a challenging and competitive place. The feeding frenzy lasts for weeks, before the remains take one last dive down through the mesopelagic zone and into darkness. In a reversal of fortunes, the benthic crabs and the octopuses will take their turn with whatever is left, picking any remaining flesh from the sturdy white bones on the seafloor.

And miles above, set upon the lurching swell, the ambergris has begun its journey.

Freshly expelled, the black and viscous ambergris—which is slightly less dense than seawater—rises slowly, ascending through the frigid ocean currents. Eventually, it reaches the surface, where it floats in the chop, forgotten and mostly submerged, sometimes for years. It can ride the swell of the southern oceans for decades. Back on land a thousand miles away, life continues. At sea, the ambergris floats: it bobs and rolls through cyclones and equatorial heat, from the Tropics to the stillness of the Doldrums, where it might be stalled for months. It picks up speed in the horse latitudes. It turns poleward and then back again. It gets trapped in ocean gyres—large rotating oceanic current systems that pieces of flotsam can spend years navigating. This journey cannot be substituted. Like wine in a bottle, ambergris slowly matures at sea. Gradually, a molecule at a time, it reacts with its surroundings until—oxidized by salt water, degraded by sunlight, and eroded by wave action—it is beached somewhere along a remote and windswept coastline much like Long

Beach; or dumped by a storm onto a busy and populated stretch of sand like Breaker Bay, in sight of a large metropolitan city like Wellington; or it washes up somewhere on the Somali coastline, or in the Chatham Islands, or the Philippines, or northern California, or on a wet little bay in Wales.

"Ambergrease is also found on the *Scots* Coasts," wrote Guy Miege in 1715, in *The Present State of Great-Britain and Ireland: In Three Parts*, "particularly on that of the Island *Bernera*, one of the *Harris* Isles, where a Weaver finding a Lump of it, and not knowing what it was, burnt it to shew him light, when the strong Scent discover'd it, and made his Head ake. It is also found on the Coasts of *Southvist*, *Kintyre* and *Orkney*."

In fact, ambergris can wash ashore anywhere there are sperm whales—which is almost everywhere, in all the world's oceans. Sperm whales are considered a "cosmopolitan" species. Unlike some other whale species, which are restricted to specific environmental bands of ocean—bowhead whales in the Arctic; Bryde's whales in the Tropics—sperm whales can be found in all the oceans at almost all but the very coldest latitudes. A sperm whale will slowly plow its way through any water, constantly diving deeply for squid, provided it is deeper than about 3,300 feet and not covered in ice. In other words, sperm whales roam almost everywhere, and some of them produce ambergris as they navigate the world's seas and oceans. The churning oceanic currents then carry the ambergris everywhere else, even to those few isolated places where sperm whales might physically be absent.

There is a randomness and unpredictability to a journey like this. It is unknowable. At various times, ambergris has been found in some strange and surprising places. In September 1908, the *Hartford Courant* reported a lucky find by a Noank, Connecticut, fisherman: while hauling up lobster pots from the bottom of Long Island Sound, John Carrington, captain of the *Ella May*, discovered that one of his traps contained a one-pound piece of ambergris. A year later the *Washington Post* described the moment that the crew of the *Hockomock*, on its return to Boston, opened up one of several swordfish taken on the Georges Bank and discovered a large piece of ambergris inside it. "The piece brought in today," the article read, "is estimated to be worth $20,000."

*   *   *

A fresh fragrant lump of ambergris could wash ashore just a handful of miles from where it was expelled. It could arrive a day or so later with the tide on Long Beach, black and sticky and smelling of fresh dung. A large valuable piece of ambergris could be there now, drying in the wind, waiting for me to find it. Or it could be carried for years instead, taken by strong currents across remote and unvisited parts of the ocean, slowly eroding until no part of it is left. There is a chance it will outlast the whale that produced it, the result of an intestinal rupture at sea. And once found, there is no way to discern the slightest information about a piece of ambergris, either when or where it was made and expelled into the ocean, or which route it took to arrive where it was found. It is simply a mystery: an artifact, a totem, a relic. Was it once part of a larger piece? Is it a year old, or has it been floating for twenty winters or more, traveling in a huge circuitous arc across the world's oceans? All but the most immediate information is unknown and cannot be discovered.

By the time an aged and well-traveled piece of ambergris arrives on the shore, though, it is different. It has been worked on by the ocean, tossed around on the waves for years like a single grain of wheat in a vast combine. Depending on how long it has been at sea, its color and texture will have evolved from a black tar-like substance to a pale, smooth waxy ball, rolling in the surf. Over the years, it loses most of its water content. It becomes smaller and denser. Its exterior hardens and takes on a tough rind-like appearance. More than anything else, it now resembles a light gray stone—a little like pumice stone, chalk, or dried clay. Its surface might have a shiny patina to it; its interior will be flecked with embedded squid beaks, like burnt black seeds. It smells pungently and, as it evolves, it undergoes another transformation: the fecal smell that characterizes freshly expelled ambergris gradually softens at sea and is replaced by a rich complex odor described variously as sweet, woody, earthy, and marine.

"Unique, illusive of precise description," wrote Robert Clarke, "the odour of ambergris has been said to suggest fine tobacco, the wood in old churches, sandalwood, the smell of the tide, fresh earth, and fresh seaweed in the sun. I myself am reminded of Brazil nuts." In an 1844 article in the *American Journal of Pharmacy*, it was said to have a "smell somewhat resembling old cowdung." An article in the *New York Times* from 1895, titled "Ambergris, the Whale Fisher's Prize," described its odor as being "like the blending of new-mown hay, the damp woodsy fragrance of a fern-copse, and the faintest possible perfume of the violet."

Whether it smells of churches, Brazil nuts, a fern copse, or all of these, it is a sought-after component of perfumes and is sold by the gram in little pebble-size pieces to independent perfumers or in bulk to those who can afford it. It is peddled in the dusty souks by herbalists in Morocco and Cairo, where it is an aphrodisiac and stirred by the teaspoon into cups of sweetened tea. Across the Middle East, it is used as incense in religious ceremonies. In China, it is eaten. Throughout history, it has been used as a medicine, as an ingredient in cooking, a component in fragrances, an adornment, a sign of wealth, an acknowledgment and celebration of the great dark unseen mystery of the ocean.

Weathered from its years spent adrift at sea, ambergris is one of the few physical manifestations of the sperm whale, an implausibly large mammal that spends most of its time miles beneath the ocean surface in complete darkness. Both literally and figuratively, any meaningful details of the journey a piece of ambergris has made are simply lost to the vastness of the deep ocean, which does not readily give up many of its secrets. The journey—which is physical, geographic, chemical, and transformative—cannot be replicated, and neither can its product. Ambergris has been synthesized, but its synthetic versions are not convincing. They lack an indefinable *something* that is gained only after years spent at sea. On completing its long journey, this nondescript sun-whitened pebble has been transformed into a prized commodity.

* * *

Back on Long Beach, I bend over to pick up another unidentified object from the sand. I examine it, smell it, and then pitch it over my shoulder, where it lands with the other rejected pieces of driftwood, seaweed, and lightweight volcanic rock. The rain is still falling. A screen of dark low thunderheads slides gracefully landward from the open sea and settles on top of the cliffs like a mantle. The tide is oceangoing. The water level drops quickly, leaving behind a fresh wet belt of flotsam that extends the length of the beach. I had read somewhere that the best time to find ambergris is immediately following a high tide—dumped by the receding waves, it sits proudly on the high-tide line, making it easily visible to anyone trying to find it. This is the time. I make sure my plastic bags are still in the pocket of my raincoat. Picking through the seaweed and driftwood, I make my way farther north, squinting into the rain. Half an hour later, I arrive at the northern end of the beach, wet and tired and empty-handed.

For several weeks, I have been calling Robert Anderson, hoping to find out more about the ambergris his son, Robbie, found on Long Beach in 2006. I leave messages on his answer machine, asking him to call me back. He never does. After a while, I am reminded of an unsuccessful few months I had spent telemarketing. Stubbornness sets in. I call three times a week. I want to know, What happened to the ambergris Robbie Anderson had found? Did the Andersons sell it? Do they still have it? And how much is it worth?

*   *   *

Robbie Anderson was lucky. But not unique. I have begun to collect accounts from the news archives of beach walkers stumbling over unusual-looking objects on beaches, taking them home, and then discovering they are ambergris. Some of the more recent of these articles carry headlines like "'MOBY SICK' FIND LANDS FRAGRANT FORTUNE" (Reuters, January 25, 2006); "GIRL IN WALES FINDS LUCKY WHALE VOMIT" (UPI, August 13, 2006); "MAN BIDS TO STRIKE RICH WITH WHALE VOMIT" (*Brisbane Times*, March 13, 2008); and "BEN STRIKES IT LUCKY WITH SMELLY FIND" (*Taranaki Daily News*, March 12, 2009).

Every time I find another account, it fills me with fresh hope that perhaps I will one day find my own ambergris. I have begun to think of it as the oddest and most intriguing substance in the world. With each new report, I drive to another local beach and walk for miles, picking up any object that I think might be ambergris and smelling it. It is the only way I know to distinguish ambergris from random debris. A couple of times a week, I walk along the tide line after a high tide, collecting all the objects that seem incongruous among the rounded pebbles and the seashells. And then, standing in the wind, I hold each one of them close to my nose and inhale, sampling the odors that cling to its still-wet surface. I have smelled several thousand rocks. I have scrutinized soft, decomposing, gelatinous pieces of kelp stalk, brightly colored fragments of plastic, black weathered pieces of coal, baskets of driftwood, lengths of twine, bleached bits of bone, pieces of beach glass, and broken oyster shells, eroded by the waves into unusual non-oyster-like shapes. One day in the rain, I almost smelled a dead wet seagull, matted and dark in the sand. But mostly I have smelled rocks, which don't smell much of anything at all.

Once, I had walked breathlessly along the shoreline after spotting a large, unusual object in the distance. It was a shoe: an old running shoe, encased in sand, unlaced by the sea. For some reason, it was a lonely and melancholy sight. At other times, I might have sat and considered the imponderable series of events that had brought this lone shoe to shore.

But not now. Clearly, this waterlogged and weather-beaten shoe was not ambergris.

For numerous reasons, smelling random objects is an imperfect technique. On several occasions, I had attracted unwelcome stares from strangers on the beach. At the rocky northern end of Long Beach one sunny afternoon, a group of climbers had watched me suspiciously until I passed out of sight behind an outcrop of rocks. More than once, I had taken my infant son with me, wearing him in a sling and fastening a rain jacket around him to shield him from the wind. To anyone else, I was a strange rotund figure who moved slowly along the beach, stooping to pick up stones from the surf, smelling them, and throwing them away. This was not normal behavior. And it was hard on the knees. Overburdened, I had fumbled several large flat rocks, almost dropping one of them on my son's head.

Clearly, another approach is needed.

2 THERE IS A PIECE AT ROME AS BIG AS A MAN'S HEAD

The best that is in the World comes from the island Mauritius; And is commonly found after a Storm. The Hogs can smell it at a great distance; who run like mad to it, and devour it commonly before the people come to it. ⋆
SIR PHILIBERTO VERNATTI, responding from Batavia, in Java, to questions put to him by members of the Royal Society of London (1667)

It is a very fruitful island, and well peopled, it produces abundance of Amber-grease, which the Inhabitants mix with their Tobacco, when they Smoak, besides that, they Sell a considerable quantity to the French. ⋆
GABRIEL DELLON, describing an island called St. Mary that lay two leagues distant from Madagascar, in *A Voyage to the East-Indies* (1698)

Inching forward in the darkness, Louis Smith has lost his bearings somewhere inside the cold and convoluted intestines of a sperm whale. In the last few minutes, he has begun to have some reservations. He has been assessing his current situation. Outside in the wet air, with the gulls racing overhead, it had seemed like a reasonable thing to do. You see a dead whale, you climb inside. Should be plenty of room in an "eight tunner," he had thought. But inside it's different. Inside, it's dark. And it's difficult to breathe. Smith rests for a second or two, takes another shallow breath, and pushes his boot heels against the thick muscular wall. But after so many days, it has grown spongy and slippery. One of his boots sinks into the tissue, and he has to shift his body, twisting awkwardly in the confined space, to tug it back out again. His ears are clogged with something. He's a little light-headed now. Reaching his arm over his head with his fingers outstretched, Louis Smith gains a few more inches. He thinks he might be upside down. Perhaps, Smith tells himself, this is like being born.

⋆ ⋆ ⋆

And so it came to pass: on July 29, 1891, Louis Smith carefully guided his boat toward the wharf at Hobart, on the island of Tasmania, carrying a strange and precious cargo. Within a couple of minutes of his arrival, Smith had caused an uproar among those gathered on the quay. According to an article in the *Hobart Mercury* newspaper, he was a fisherman—"a man of colour, from Recherche," a small town on the southern coast of Tasmania. He was a simple man. But that was about to change. Sitting in the stern of Smith's boat, covered with sacking and tarpaulin, was an enormous boulder of ambergris. It weighed an estimated 180 pounds. Smith obliged the onlookers, pulling aside the tarpaulin so that they could see the ambergris wrapped within its folds.

"The piece is, roughly speaking, about 22 in square," read the *Mercury* article, "smells somewhat like guano, and resembles sepia in colour." A loud murmur ran the length of the wharf as those present speculated on its value. There were disagreements. Among those in the crowd, some said it wasn't even ambergris. Others shouted out offers to buy it on the spot. "Old whalers, and those who professed to know, estimated it as worth ten thousand pounds, and some thought it as many shillings."

Smith was known locally as Black Louis. He had been looking for ambergris for a long time, he told reporters, but had rarely found much of it, and never a piece as large as the boulder that sat in the back of his boat. So far, Smith's story was unusual but not unique. Not yet. It only became unique—and odd, and unsettling, and stomach-churning—when he began to tell the crowd where he had found the ambergris. It came, he told them, from a sperm whale that had been captured and killed ten days earlier by a ship called the *Waterwitch*. "It was about an 'eight tunner,'" Smith said. Its blubber had been cut away and "tried out," or boiled down to oil, in a large iron try pot. Afterward, the whale had been unceremoniously dumped in the ocean, and Smith had seen it floating there. Without thinking, he hitched the carcass, stripped of its blubber, to his boat and towed it home to Recherche. There, with the gulls wheeling overhead, Smith had stood on the shore next to the cold dead whale and wondered if there might still be some ambergris inside it.

Working quickly, he took out his knife and cut a large hole in the throat of the ten-day-old carcass. And then he peered inside. Blackness. Nothing. He inserted his outstretched arms and his head into the hole, and then began slowly rolling his shoulders around until he'd worked those in too. And then he crawled inside. *He crawled inside the whale.* Once inside, Smith pushed and corkscrewed and slipped his way through its slowly disintegrating intestines until he found the boulder of ambergris that was stuck there.

*   *   *

If Louis Smith's story tells us anything, it's that at one time in history, and not so very long ago, people were willing to go to great lengths to obtain

Some excitement was caused on the wharf (says the *Hobart Mercury*) when it became known that a fisherman named Louis Smith, a man of colour, from Recherche, had brought to town between 180lb and 200lb of ambergris. The treasure was in the stern of Smith's boat, carefully covered with sacking and tarpaulin. There certainly was not very much to look at when the coverings were lifted, but speculation ran high on the wharf as to its true value. Old whalers, and those who professed to know, estimated it as worth £10,000, and some thought it as many shillings. The piece is, roughly speaking, about 22in square, smells somewhat like guano, and resembles sepia in colour. Smith says that the whale from which he obtained it was about an 'eight tunner,' and had been captured and 'tried out' by the Waterwitch on Sunday, the 19th inst., and that it was of the spermaceti kind. To obtain the ambergris Smith had to cut a large hole in the throat and make his way to that part of the whale's intestines where such deposits are usually found. He says that very few whales contain it. He has sought for it a long time and has rarely been rewarded. Several offers were made on the spot for the ambergris as it stood, but Smith firmly declined them all, and eventually procured a strong box, in which it was deposited and sealed, and placed in the keeping of the Commercial Bank

"A find of ambergris," *Timaru Herald* 53, no. 5212 (August 11, 1891), downloaded from Papers Past, with permission of the Alexander Turnbull Library, Wellington, New Zealand.

ambergris. But what is ambergris, and what does it do? The answers to these questions are both simple and not simple at all. Much about it remains a mystery. It is still not known, for instance, whether whales die passing ambergris or expel it and swim on, relieved to be lighter. Ambergris has a strange and hard-to-describe smell. It floats in the ocean and then washes ashore randomly, unpredictably, often after heavy storms and rough seas. It is so rare that it can be found almost anywhere, but is hardly ever found anywhere at all. In fact, for a long period in human history, no one even knew where ambergris came from, what it was, or how it was made. But when made into a tincture—ground into a fine powder and dissolved in alcohol—and added to perfume, ambergris acts as a powerful fixative, slowing the breakdown of a fragrance on the skin and making the scent last longer on the wearer. It is impossible to know when this strange and esoteric discovery was made, who made it, or how.

But, for this most important reason, ambergris was a singular and valuable substance throughout history. It was worth as much as gold. At different times, depending on the global supplies of ambergris, it was worth *more* than gold—twice as much, or three times. In the twentieth century, chemists synthesized ambergris. The chemical analogue is not convincing, and ambergris is still traded and used around the world. At one time, it was known in many places as "floating gold." It was prized by monarchs and came from remote and far-flung places: the Nicobar Islands and the Molucca Islands; Sumatra, Japan, and the Andaman Islands; and Somalia and Ethiopia.

"Ambergris," wrote Caspar Neumann in 1729, in the *Philosophical Transactions and Collections of the Royal Society of London*, "is mostly brought from the East-Indies, from and about the islands of Madagascar, Molucca, St. Maurice, Sumatra, Borneo, Cape Commorin at Malabar, and from the Ethiopian shoars, which are said to produce Ambergris from Sofala quite to Brana. Besides as Ambergris is carried to great Distances by the Sea, there are a hundred other Places in the World, where it may be found."

In many of these places, ambergris became an important commodity. In the Nicobar Islands, beginning as early as the ninth century, ambergris was harvested from the coastline and traded with passing ships for iron. In the early months of 1688, William Dampier was aboard the *Cygnet*, a ship filled with violent and sea-hardened criminals. Dampier was an English navigator and a buccaneer. A seafaring man, he had spent his entire career on various different ships and was the first man to circumnavigate the world three times. But in 1688, on board the *Cygnet*, the crew had grown restless. A year earlier, it had enjoyed what the *Encyclopaedia Britannica* (1910) called "six months' drunkenness and debauchery in the Philippines." After that, the crew had mutinied, leaving behind their captain, a man they had, in harder times, planned to kill and eat. He was, Dampier had already noted in his *Journals*, "lusty and fleshy." Since mutineering, they had meandered across the ocean, tracing an irregular route: first west, from the Philippines to the Chinese coast, and then back again to make another circuit of the Philippines, then changing course yet again and passing quickly through the Spice Islands, before stopping briefly in Australia and turning toward Sumatra.

Eventually, from Sumatra, they sailed to the Nicobar Islands. On May 5, 1688, at ten o'clock in the morning, the *Cygnet* was finally anchored at the northwest end of Nicobar Island "in a small Bay, in 8 Fathom Water, not half a Mile from the Shore."

In Dampier's published *Journals*—his portrait, with his doleful down-cast black eyes, on the frontispiece—he wrote:

> Ambergrease is often found by the Native Indians of these Islands, who know it very well; as also know how to cheat ignorant Strangers with a certain mixture like it. Several of our Men bought such of them for a small Purchase. Captain Weldon also about this time touched at some of these Islands, to the North of the Island where we lay; and I saw a great deal of such Ambergrease, that one of his Men bought there; but it was not good, having no smell at all. Yet I saw some there very good and fragrant.

By this time, Dampier had lost his appetite for plunder. He had been at sea for most of the previous ten years—buccaneering in the South Seas, West Indies, and the American colonies. He asked to be let off the ship, along with two other Englishmen, a Portuguese man, and some Malay sailors. Dampier explained:

> Indeed one reason that put me on the thoughts of staying at this particular place, besides the present opportunity of leaving Captain Read, which I did always intend to do, as soon as I could, was that I had here a prospect of advancing a profitable Trade for Ambergrease with these People, and of gaining a considerable Fortune to my self; For in a short time I might have learned their Language, and by accustoming myself to row with them in the Proes or Canoas, especially by conforming myself to their Customs and Manners of Living, I should have seen how they got their Ambergrease, and have known what quantities they get, and the time of the Year when most is found.

He was unsuccessful. The Nicobarese maintained control of their local ambergris trade. And two hundred years later, they still had control of it. The following is taken from an 1870 report written for the India Home Office, titled *Papers Relating to the Nicobar Islands*:

> Ambergris is found in all the group of the Nicobars; and some years in such quantities that this article is scarcely of any value in these islands. In the various islands I visited, the natives brought me ambergris for sale; but its having been mixed with the wax of a small bee, which establishes itself in the trunk of decayed trees, it was of a very inferior quality. The genuine ambergris is sold very dear at Penang. The Chinese and Burmese use it for medicinal purposes.

*   *   *

Included in the *Papers Relating to the Nicobar Islands* is a limited vocabulary of Nancowry, a separate dialect of the Nicobarese language. The reproduced list is fewer than forty words long—and most of them are commonly used nouns like *man, woman, eyes,* and *house.* Nevertheless, the Nancowry word for *ambergris* is listed. If you're ever in the Nicobar Islands and someone tries to offer you inferior ambergris, which has been mixed with the wax of a small bee, you'll need to know what that word is. And so this is it: *Kampei.*

A *Dictionary, Hindoostanee and English* from 1820 has several entries for ambergris, called *umbur,* and the following is a term for hair perfumed with ambergris: *zoolfi umbureen.* From 1884, *A Dictionary of Urdu, Classical Hindi, and English* includes the term *ambar-ca*: an ornament for the neck, full of ambergris. And the entry for *cowd*: an unguent or fragrant paste of four ingredients (ambergris, saffron, mush, and the juice of the flowers of the *Arbor-tristis*). A quick look at *A Comprehensive Persian-English Dictionary* from 1992 proves how important ambergris was to the Persians. There are numerous ambergris-related entries. For instance, the term *sara* means "pure" when used to describe gold, ambergris, or musk. *Sara* is also "the name of a place on the coast of Oman, celebrated for its ambergris." The words *ambar-shamim* are used to describe a rose as fragrant as ambergris; and a perfume compounded of ambergris, musk, and wood of aloes is known as *ambarina.* Stranger still, the Persian word for cow is *gawi,* and a manatee is a *gawi-ambar*—or ambergris cow—a name that stemmed from the misguided belief that manatees produced ambergris.

The Chinese believed ambergris was dragon's spittle that had fallen into the sea and solidified. By the sixteenth century, they were calling it *lung sien hiang,* which in Chinese means "dragon's spittle fragrance." In Japan it was known as *kujira no fu,* or whale dung. Across much of the Middle East, it was known as *anbar.* In Europe, its resemblance to Baltic amber—or fossilized tree sap—earned it the name *ambergris,* a contraction of the French for gray amber. Elsewhere, and at other times, it was referred to as *ambergreen, ambergreece, ambergrease,* or simply *amber.*

In 1671, when John Milton mentioned ambergris in *Paradise Regained,* he called it *grisamber,* describing a meal that included "fowl of game, in pastry built, or from the spit, or boiled, grisamber steamed." An annotated edition of *Paradise Regained* from 1753 has the following footnote:

> A curious lady communicated the following remarks upon this passage to Mr. Peck, which we will here transcribe: "Grey amber is the amber our author here speaks of, and it melts like butter. It was formerly a main ingredient in every concert for a banquet; *viz.* to fume the meat with, and that whether boiled, roasted, or baked; laid often on the top of a baked pudding; which last

I have eat of at an old courtier's table. And I remember, in an old chronicle there is much complaint of the nobilities being made sick at Cardinal Wolsey's banquets, with rich scented cates and dishes most costly dressed with ambergris. I also recollect I once saw a little book writ by a gentlewoman of Queen Elizabeth's court, where ambergris is mention'd as the haut-gout of that age."

*   *   *

Some of the earliest written references to ambergris date back to 700 AD. In approximately 947 AD or so, the Arab geographer al-Masudi completed his monumental work *Meadows of Gold and Mines of Gems*, a recounting of his travel to places like India, Zanzibar, Oman, the Caspian Sea, and elsewhere. Al-Masudi was known as the Herodotus of the Arabs. In his writings, he mentioned ambergris numerous times: it was found on the Syrian coastline; and ambergris that washed ashore along the Spanish coastline was exported to Egypt. While traveling though a region of southern Arabia, al-Masudi wrote:

> They are a poor and needy people: they have a sort of camel called Mahri camel: it goes as fast as the Bejawi camel, or even faster, as some think. On these they ride along their coast; and when the camel comes to ambergris, which has been thrown out by the sea, it kneels down; for it is trained and taught to do so: thus the rider can pick it up. The ambergris, which is found on this coast, and on the inlands and coast of el-Zanj, is the best: it is round, of a blue colour, and is the size of an ostrich's egg, or smaller.

By the time Louis Smith harvested his ambergris, people had been risking their lives to collect and trade it for more than a thousand years. In fact, by the ninth century, Persian merchants were traveling as far afield as Somalia and Kenya, on the east coast of Africa, to trade with natives for the ambergris that washed ashore there. It was Persian merchants, too, who later traded iron for ambergris in the Nicobar Islands. By the thirteenth century, the Chinese had clearly entered the ambergris trade. Between 1405 and 1433, the ships of the Ming Dynasty treasure fleet traveled as far afield as Vietnam, Sri Lanka, the Persian Gulf, and Africa, collecting precious stones, spices, and other trade goods, including ambergris.

From the bustling ports of Persia and China, ambergris was transported along the trade routes. Fragrant mottled pieces of ambergris, gathered from places like Borneo, Somalia, and the Nicobar Islands, slowly moved west toward Venice and western Europe, alongside ginger from Malabar, rhubarb from Persia, and syruped fruits from Palermo.

* * *

When we had the time, my wife and I visited local beaches, strapping our son into a stroller and bouncing him across the wet sand at low tide. We sifted through tangled plaits of kelp for the smallest nugget of ambergris, filling plastic bags with pinecones, eroded bars of soap, and oyster shells. I walked to the local grocery store and pinned a handwritten note to the communal notice board: "*Have You Found Ambergris?*" For a week, it sat between a sign that read, in sloppy handwriting, "*For Sale: 2-yr-old goat. Good eating!! $150*," and another with the stark headline: "*Oar Found—At Deborah Bay.*" I heard nothing.

In the evenings, I worked through dozens of impenetrable texts, mostly written in the seventeenth and eighteenth centuries, which contained obscure references to ambergris. The following passage from Caspar Neumann's 1729 dissertation in *Philosophical Transactions* is so enjoyable, I'm reproducing it in full:

> It is worthy of our Consideration, that this precious Bitumen is frequently found in *very large Masses*. I will not insist on what *Faber Lyncaeus* relates from *Gregory de Bolivar*, that there are Pieces of Ambergris found, weighing 100,000 lb. much less what is extant in *Gracias ab Horto*, that there are whole Islands full of Ambergris, much less shall I regard what is told by one *Isaac Vigny*, a *Frenchman*, who had traveled, that he knows a Country, so rich in Ambergris, that a hundred ships might be laden with it. These I say are mere hyperbolic Fictions; but the following are credible, or may serve at least to prove the Certainty of great Masses of Ambergris being found. In 1555, at *Cape Commorin*, a Piece of Ambergris of about 3000 lb. was found, and sold at that Time for Asphaltus, or common Bitumen. *Joh. Hugo Lindotsch* says, there was a Piece formerly found about this *Cape* weighing 30 Quintals. *Monardes* and *Hernandez* mention Pieces of 100 lb. *Gracias ab Horto* mentions one that was of the bigness of a Man, and another that was 90 Hands breadth in Length, and 18 in Breadth. *Montanus* speaks of a Piece of 130 lb. which was kept by the King in *Satsuma*, in 1659. In 1666, a Piece was thrown up at the River *Gambia* near *Cape Verde*, that weighed 80 lb. and was brought into *Holland*. In 1691 there was a Mass of 42 lb. at Amsterdam. *Daniel de Bruel* affirms, that a Piece was found about *Malacca* of 33 lb. There is a Piece at *Rome* as big as a Man's head. Both at *Rome* and at *Loretto*, and in many other places of *Italy*, there are many Curiosities artificially made of Ambergris, which evidently appear to

have been made out of very large Pieces. The above-mentioned *Vigny* brought a considerable Piece from the *East-Indies*, for he sold it for 1300 *l.* Sterling. *Kaempfer* also testifies, that in his Time a Piece was found in *Japan*, weighing 100 Catti, or about 130 *Dutch* Pounds. The two Brothers *Joh. Andreas* and *Marcus Matsperger*, in 1613 bought a Piece of *Robert Struzzi* at *Venice*, weighing 48 lb. 8 oz. But to mention no more, we have a late and most convincing Example in that great Piece of Ambergris, which the Dutch *East-India* Company bought of the King of *Tidore* for 11,000 dollars. It was at first of the Shape of a Tortoise, weighed 182 lb. was 5 Feet 8 Inches thick, and 2 Feet 2 Inches long. *Chevalier* has given a prolix Description of it in a little Treatise printed in *Amsterdam* in 1700, and had added various Figures representing it in different views. It was kept many Years at *Amsterdam*, and after it had been shewn as a great Rarity to several hundred, perhaps thousands, Persons; was at last broken to Pieces, and sold by Auction, so that many Persons now alive have been witnesses to it, and consequently it can no longer be doubted that Ambergris is found in very large Masses.

I'm not sure what I find so appealing about this passage. It seems like such an odd combination of erudition, gossip, and plain strangeness: the fact, for example, that the king of Tidore had a tortoise-shaped piece of ambergris measuring more than five feet long is something I never knew I needed to know. But, somehow, I feel better for knowing it. The discovery that there were numerous curiosities—in Rome and Loretto, and elsewhere in Italy—that were painstakingly whittled from larger pieces of ambergris is also a revelation. Where are these curiosities now? What happened to them? Is there still a statue hidden in a dusty attic somewhere in Rome? Perhaps a little fragrant statue of Romulus and Remus, fashioned from 300-year-old ambergris?

*   *   *

After sailing through oceans crowded with whales, early settlers to North America understood immediately that they had found a source of whale oil and baleen, or whalebone. By the first half of the eighteenth century, the American whaling era had begun. Ambergris was just another lucrative product to be harvested from sperm whales.

"In Asia, and part of Africa," wrote Henry William Dewhurst in *The Natural History of the Order Cetacea, and the Oceanic Inhabitants of the Arctic Regions* in 1834, "ambergrease is not only used in medicine, and as a perfume, but considerable use is also made of it in cookery, by adding it to several dishes as a spice. A great quantity of it is constantly bought by the pilgrims who travel to Mecca, who probably offer it there for the purpose of incense; in the same way that frankincense is used by the clergy in the performance of the sacred ceremonies of the Roman Catholic church."

Herman Melville's *Moby-Dick* remains the finest literary representation of the American whaling era. Published in 1851, Melville devoted an entire chapter of *Moby-Dick*—and one of the most memorable scenes in the book—to ambergris. One day, while sailing on calm seas, the *Pequod* draws up alongside another ship called the *Rose-Bud*, which is towing behind it two decaying whale carcasses. Stubb, the first mate of the *Pequod*, convinces the French captain of the *Rose-Bud* to unhitch the whales from his ship, claiming that several crew members of another ship had recently died, killed by a cloud of poisonous gases escaping from whale carcasses.

The captain follows his suggestion, and once he is gone, Stubb clambers on top of the smaller and thinner of the two whales: "Seizing his sharp boat-spade," wrote Melville, "he commenced an evacuation in the body, a little behind the side fin. You would have almost thought he was digging a cellar there in the sea; and when at length his spade struck against the gaunt ribs, it was like turning up old Roman tiles and pottery buried in fat English loam."

Stubb was searching for ambergris, digging into the carcass at precisely the point that he knew it would be found.

> "I have it, I have it," cried Stubb, with delight, striking something in the subterranean regions, "a purse! a purse!" Dropping his spade, he thrust both hands in, and drew out handfuls of something that looked like ripe Windsor soap, or rich mottled old cheese; very unctuous and savory withal. You might easily dent it with your thumb; it is of a hue between yellow and ash color. And this, good friends, is ambergris, worth a gold guinea an ounce to any druggist. Some six handfuls were obtained; but more was unavoidably lost in the sea, and still more, perhaps, might have been secured were it not for impatient Ahab's loud command to Stubb to desist, and come on board, else the ship would bid them goodbye.

A whaler as seasoned as Stubb would have known exactly where to look for ambergris and how to recognize it. A whaling cruise could last as long as five years—five grueling winters spent at sea, often in the challenging conditions of the frigid Antarctic oceans. If a large greasy boulder was harvested on the flensing deck, hauled from the slippery innards of a dead sperm whale, it could be worth as much as all the whale oil collected during the rest of the voyage. Such was the case in 1858, when the

Watchman returned to Nantucket after a yearlong voyage: stowed away with the other cargo were four casks filled with a total of more than eight hundred pounds of ambergris.

* * *

At one time, ambergris had been ubiquitous. It was traded like gold, prized everywhere from the east coast of Africa to Alaska, bartered and quarreled over, and transported from remote places like Tasmania and the Andaman Islands, to modern cities—to perfume houses in Paris and London—on the other side of the world. And it was used, at one time or another, for almost every imaginable purpose.

The natives of Sulu, an independent island province near the Philippines, had once burned large lumps of it, fishing by the light it provided on dark and moonless nights. And in Mexico, Moctezuma, the Aztec emperor (1397–1469), supposedly added ambergris to his tobacco. During the Middle Ages, the English believed that a piece of ambergris, held tightly beneath the nose, would protect them from pestilence. Lazare Rivière, a seventeenth-century French physician, reported that ambergris was an effective cure for rabies. When the governor of Mozambique finished his three-year term, Jean-Baptiste Tavernier reported in 1678 in *Travels in India*, he was partly paid in ambergris, taking home "300,000 pardos' worth of ambergris, and the pardo, as I have elsewhere said, amounts to 27 sols of our money." The Florida Native Americans, and the Bermudians and the Bahamians too, valued the medicinal properties of ambergris. "Whenever they are poisoned with fish," wrote Henry Barham in 1794, in *Hortus Americanus*, "(which often they are), they fly to ambergris as a powerful antidote, and are cured therewith, and rescued from the most horrid symptoms threatening them."

In Asia and the Middle East, the effects of ambergris also were widely celebrated. "The belief in its efficacy yet lingers in the Orient," stated an article about ambergris in *Arthur's Illustrated Home Magazine* from 1874, "where it forms the chief ingredient in a very popular so-called 'Elixir of Life.' In Egypt it is valued principally for its supposed virtues as an exciter of love." The most common method of preparation in Egypt involved melting a carat-weight of ambergris in the bottom of a coffeepot before making the coffee. Sailors simply ate crude lumps of it. In the mid-nineteenth century, the standard chemical formula for Indian ink included a few drops of essence of ambergris.

"It is a signal remedy for the horrid spasms, or loss of the use of limbs in the dry belly-ache," wrote Barham in 1794. He continued: "It also stops vomiting and looosenesses, is proper for all inward bruises, and a most universal cordial; it refreshes the memory, and eases all pains to the head, being dissolved in a warm mortar and mixed with ointment of orange-flowers, anointing the head, temples, and forehead therewith; it also helpeth barrenness proceeding from a cold cause, and cures fits of the mother inwardly taken."

After reading such an enthusiastic testimonial, I had begun to consider eating ambergris too, if I ever found any.

* * *

By August 1891, Louis Smith's ambergris had arrived in London—an implausibly large boulder. Rumors of its arrival were beginning to spread through the city. It was heavy, cumbersome, and difficult to move. Attempts to place it inside a train car and to transport it to potential brokers had failed. It was too large to move. Instead, it was stored long-term—sealed in an airtight case—and locked in a London bank vault. It had earned a nickname: the Bank Lot.

Eventually, brokers were summoned to inspect the Bank Lot. When the case was opened, "everyone beat a hasty retreat from its vicinity, for the horrible smell which issued from the box was overpowering." Unable at first to approach the case, they waited for the smell to subside before unpacking the large black object inside. Finally, it was lifted from the case and placed before the assembled bank officials and brokers: the Bank Lot measured six feet four inches in circumference and was almost perfectly spherical. It weighed 162 pounds 11 ounces. It was simply an enormous black boulder of ambergris.

Included in the 1902 U.S. Commission of Fish and Fisheries' *Report of the Commissioner* was an excerpt of a letter written by one of the brokers: "The next thing to do was to split the lump, so as to see what the interior was like. This was accomplished with the aid of long chisels and crowbars. We then saw that the substance consisted of layers or laminae, rolled around a central core, the laminae varying a good deal in texture, color, and flavor."

There, in the stillness of the vault, the Bank Lot was carefully dismantled. The core was removed from the surrounding layers. Shaped like a rifle bullet, it weighed five pounds and "stood alone, a pure, solid lump of the finest gray ambergris."

People began to talk about the Bank Lot. It was inevitable. An enormous and priceless black boulder of ambergris, brought all the way from Tasmania, was now hidden somewhere within a bank vault in London. The brokers knew that no one would be able to buy so much ambergris in a single lot. In the summer of 1891, the price of ambergris was unusually high—108 shillings per ounce. If the details of such a gigantic boulder of ambergris became known, market prices would plummet. The brokers agreed on a strategy of deception and counter-deception. If anyone asked them about the Bank Lot, they said it was a myth.

From an article in the trade journal *Chemist and Druggist*: "Our representative, who called for information at the bank, was confronted with a number of courteous officials whose know-nothing attitude would have baffled a Sherlock Holmes." The subterfuge had begun. Referring to the Bank Lot as the Monster Lump, the article continued, "We came into contact with several gentlemen, each of whom assured us, in strict confidence, that the lump had been entrusted for sale to his care, and that fabulous prices had already been obtained for part of it."

Eventually, ambergris always disappeared. It was impossible to track. Found somewhere on a remote beach—maybe in New Zealand, or at the steaming edge of a lush frond-filled jungle in the Nicobar Islands, or even inside the intestines of a decaying sperm whale—it was then sold, passed along from vendor to vendor, and transported to London or Paris. And then, it simply disappeared. It was processed: ground down with sand in a mortar, dissolved in alcohol over a period of months, made into a tincture, incorporated into perfume, and then sprayed onto the slender necks and the blue-veined wrists of the wealthy. But the Bank Lot was different. It was such a superlatively large and singular piece of ambergris that it left a trail of rumor and excitement wherever it went. Press reports followed it from Hobart, to Melbourne, and finally to London, where it sat in its bank vault.

"The recent importation of a piece of ambergris from Melbourne, weighing, it is said, 136 pounds and valued at £10,000, naturally caused a good deal of excitement," reported the *American Druggist* from London in 1891. "The piece is believed to be the same which was captured by a black man in Tasmania some time ago. But the matter still remains shrouded in some mystery, for the London consignees of the parcel refuse to show the piece to anyone, and even decline to give the slightest information of any value."

Even a glittering black boulder of ambergris like the Bank Lot disappears eventually. In London, the brokers quietly sold it off, releasing a piece at a time to avoid flooding the market, until it was gone.

"The fact weighed heavily upon us," wrote the broker, in his 1902 letter to the U.S. Commission of Fish and Fisheries,

> that if the real truth about [the ambergris] leaked the depression of the market would be so great that we should not be able to do justice to our clients, and, consequently, as few people as possible were let into the secret. It is true that reports about it were rife for a month or two, but as nothing authentic could be ascertained they gradually died out, and we have ourselves been repeatedly assured that the thing was a myth altogether, one gentleman going so far as to tell one of our partners, about three months afterwards, that he held three-fourths of the total quantity of ambergris in London, not knowing that we were controlling about one-and-a-half hundredweight.

For the few odd minutes in July 1891 spent clambering through the whale—what amounted, more or less, to a brief and slippery second birth—Louis Smith earned £11,000. It was an astonishing and transformative amount of money.

"It is a matter of some regret to us," the broker admitted sadly, "that we did not secure a photograph of this extraordinary lump."

3 THE BEACH MAFIA

I thought it was just an ordinary floating rock. ∗ BEN MARSH, to
reporters from the *Taranaki Daily News* (March 2009)

"I'm not really interested in discussing it with you to be honest, mate,"
Rob Anderson says abruptly on the phone.

I called Anderson to ask him about his son Robbie's discovery of ambergris—reportedly worth more than $10,000—on Long Beach in May
2006. Long Beach was just a few miles from my house. I visited it often.
I wanted to learn more about the chunk of ambergris Robbie had found
on the gently sloping sand there three years earlier. Several times in previous weeks, I had called and left messages, which Anderson had never
returned.

"I'm just not interested," he says again. "That's why I didn't get back
to you."

I ask Anderson if the object his son had found was even ambergris,
and if it had been sold.

"Oh, it's ambergris," he answers. "It hasn't been sold because, as I explained to my son, who has since found more, it will only increase in value.
It won't decrease. When you don't need the money, you hold on to it."

I say nothing. Anderson says nothing. The seconds pass. "Apparently,
it's been appraised at $15,000," he admits finally, and then, as if regretting
the disclosure, he adds, "or whatever."

A moment later, Rob Anderson hangs up on me. I feel, and not for
the first time, like a failed telemarketer. From my window, I can see the
ruffled blue water of Otago Harbor. The hills of the peninsula rise up
sharply behind it like a grassy green wall, volcanic and steep, plunging
beneath the surface of the water, angling downward to give the harbor its
depth. Farther to the north a few miles sits Long Beach and the gleaming
breakers of the Pacific Ocean.

For days afterward, I struggle to understand why my phone call with

Rob Anderson ended the way it did, with me holding a dead phone. When Robbie had first returned from the windblown strip of sand near their home, holding a two-pound lump of ambergris, news reporters had found his father approachable enough. What had caused him to grow so tight-lipped since then?

And then three words from our brief conversation slowly filter back to me: *Since found more.*

I imagine Rob Anderson and his son, Robbie, doubled over on Long Beach as the slate-gray rain clouds drift across the sky above them. *Since found more.* Perhaps they scour the sand after every high tide, patiently sifting through the debris for the smallest fragments of ambergris. And maybe at home, a pile of ambergris is slowly accumulating in a corner of their kitchen, one waxy, fragrant piece at a time.

* * *

One thing had become abundantly clear: I needed to speak with people who knew more about ambergris than I did. For more than six months, that was what I attempted to do. And without exception, I failed.

The first thing I learned is that mentioning ambergris can change a conversation. Talkative people suddenly become suspicious and non-communicative. Only those who know nothing about ambergris will discuss it in any detail. Aware of the impact of revealing too much, anyone familiar with its value and scarcity will say nothing or respond in generalities to questions that require specific answers. A pattern began to develop: as I carefully explored each new line of communication, I was referred to a succession of steadily less helpful people, each inclined to say fewer words than the last, until the trail finally ended. Requests to meet went unanswered. Telephone numbers rang out endlessly when I dialed them. The people I expected to know about ambergris always knew nothing; and the people who actually *did* know something refused to talk with me.

I began my search for information at the local aquarium, a cluster of buildings perched on a wild green finger of land that protrudes into the channel of Otago Harbor. Program director Sally Carson referred me to Steve Dawson, a University of Otago associate professor of marine studies and a trustee of the New Zealand Whale and Dolphin Trust. A few days later, Dawson responded and admitted he had no experience with ambergris, suggesting I contact Dallas Bradley, a professional ambergris trader based in Invercargill, a three-hour drive south of Dunedin.

Dawson gave me Bradley's mailing address. There was no phone number or any other information, just an address. Undaunted, I found a post-card with a photograph of a whale on it, wrote a quick note on the back along with my contact details, and sent it to Bradley in Invercargill. It would arrive the following day. I had the distinct feeling that I might as well have driven over the hills to Long Beach, stood on the sand beneath the cliffs, and thrown the postcard into the rolling surf instead. I waited for a few days. And then a week passed. And a month. No response.

A few weeks later, I contacted Hal Whitehead, a world expert on sperm whales and a professor at Dalhousie University in Nova Scotia, Canada. In 2003 Whitehead published *Sperm Whales: Social Evolution in the Ocean*. Despite years spent researching sperm whales from his oceangoing re-search vessel, the *Balaena*, Whitehead responded that his research inter-est is in sperm whale social behavior, and he wouldn't be able to provide any information about ambergris. I called Richard Sabin, the curator of marine mammals for the Natural History Museum in London, who re-plied: "I would not be able to give you much more information than can be found in the scientific literature."

Next, I contacted Luca Turin, a fragrance chemist who has written sev-eral books on fragrance, including *The Secret of Scent: Adventures in Per-fume and the Science of Smell*. Turin was friendly and graciously referred me to Christopher Sheldrake, an expert on natural and synthetic com-pounds for Chanel in Neuilly, near Paris, who ignored my e-mails and phone calls. Around this time, I also contacted Dale Rice. Recently retired from the National Oceanic and Atmospheric Administration laboratory, Rice had written a brief but informative entry for ambergris in the *Ency-clopedia of Marine Mammals* in 2009. He returned my e-mail, suggesting that I contact Robert Clarke, the author of "The Origin of Ambergris."

On another occasion, a bright and cheerful e-mail correspondence with an Auckland-based ambergris vendor became surly when I asked one too many questions. I requested more specific information about the largest shipments of ambergris she had bought and sold, both in terms of weight and value, and she responded: "No, sorry I can't help you with this information. We run a business, we have competitors and this is commercially sensitive information." There the correspondence ended, abruptly and permanently.

* * *

On October 21, 2003, Ross Sherman was fishing a few miles north of Bay-lys Beach, not too far from Dargaville on New Zealand's North Island, when he noticed a car racing toward him across the sand. Among the several different competing versions of what happened next, many details are disputed. But this is not one of them. It's a filmic moment: a car plowing across the wet sand toward Ross Sherman. From that moment on, almost everything else becomes contentious. One other important detail is not disputed: John James Vodanovich was the driver of the car, as it slid and fishtailed along the shoreline toward Sherman.

Three years earlier, the two men—Sherman on the beach; Vodanovich in the bouncing car—had been business partners in a local ambergris-collecting venture. It was, more or less, a full-time concern for both men. Daily they harvested the beaches around Dargaville for ambergris, searching the beach after every high tide. It was a practice they called "milking the beach." Eventually, the partnership ended badly. The two men hadn't spoken since.

"It finished badly for him, not for me," Vodanovich later corrected the Whangarei District Court.

Back on the beach, the car was still speeding along the shoreline toward Sherman, a brown spray of sand in its wake. In a modern interpretation of a medieval joust, Crown prosecutors argued that on that day in October 2003 Vodanovich, a self-employed seaweed gatherer, had intentionally aimed his car at Sherman on the beach. In self-defense, Sherman had allegedly swung a length of plastic piping at the car. The car had hit him. He was thrown in the air and landed heavily on the beach. Consequently, Vodanovich was appearing in court to face several charges relating to the collision. After he hit Sherman, Vodanovich drove his damaged and dented car home. "I just drove off," he said in his police statement, adding that he had seen Sherman running across the beach toward him, brandishing a length of pipe. "I feared for my life," Vodanovich had said. "I thought he was probably lining me up for another shot."

Meanwhile, Sherman was lying flat on the sand listening to the sound of Vodanovich's car recede into the distance. Once the beach was quiet again, Sherman struggled to his feet, staggered to his car, and drove to a local police station. In court, a doctor testified that the injuries to Sherman's hands, feet, and legs were consistent with being struck by a motor vehicle. When he arrived at the police station, the *Northern Advocate* reported, Sherman "had tooted his vehicle's horn to get the attention of police inside the station, as he believed his injuries meant he could not get out of the vehicle."

Almost a year later, Vodanovich was found not guilty of both charges.

It had been Sherman's word against Vodanovich's. A headline in the *New Zealand Herald* cheerily announced the news: "NORTHLAND MAN CLEARED OF USING CAR AS WEAPON." The case was over. A fragile calm descended over the Dargaville coastline.

And it lasted. At least for a little while.

As I read about the battles that were fought over ambergris in 2003, I thought of what Rob Anderson had said again: *Since found more.* And I began to understand what had made him so reluctant to talk. Perhaps it was the fear that unwanted incidents—like the one between Vodanovich and Sherman—might take place on the unpopulated sands of Long Beach.

* * *

It's a quiet Sunday morning. Early summer. Out on the coast, a chill still clings to the pearly air. I trudge slowly around the broad, kelp-choked bay at Brighton, a few miles south of Dunedin. The tide is oceangoing, and I walk the meandering high-tide line. In an attempt to avoid the dog walkers and the weekend strollers, I keep to the less visited parts of the bay. I pick up a long wet piece of driftwood, covered with thick tufts of bright-green intestine seaweed, which I use to give objects on the shoreline an exploratory poke. When I find wet tangles of seaweed and heavy pieces of weathered timber, I use the driftwood like a crowbar, digging it into the sand and prying them into the air so that I can peer into the dark spaces beneath them. Earlier that day, I had driven past sheep farms in the gray morning light, following the coastline south. The road is bordered by a rolling green apron of land. Just beyond it, the Pacific Ocean glitters like a flat blue table. Pulling into the parking lot in my old and dented Subaru, I realized I still had no idea what I was looking for, or how I should try to find it.

At home earlier that morning, I had read a December 2006 article in the *New York Times* with a headline that reminded me of a 1950s B movie: "PLEASE LET IT BE WHALE VOMIT, NOT JUST SEA JUNK." It described the circumstances by which Dorothy Ferreira, a sixty-seven-year-old woman from Long Island, became the owner of a strange green object that she hoped was ambergris. In an accompanying photograph, Ferreira—a gray helmet of hair, a smudge of bright red lipstick—peered myopically from behind the object, gripping it tightly in both hands like a steering wheel. Against the wood paneling in Ferreira's home, it was a translucent, sickly olive-green color. Its dimpled surface was an accumulation, a confusion really, of long, thin tangles, which reminded me of

worm castings—the intricate little piles of mud work sometimes left behind on a lawn by earthworms that have tunneled beneath it.

Several days earlier, Ferreira had received a box from her eighty-
two-year-old sister in Iowa. Inside the box: the object. It weighed four
pounds. Her sister had found it in Fort Pond Bay, off the Long Island
Sound, in the mid-1950s. Decades later she moved to Iowa and took it
with her: a bright green, heart-shaped tangle of unidentified material she
had found while beachcombing with her dog, half her lifetime ago. For
more than fifty years, she had wondered what the object was and whether
it was valuable. When asked by the *Times* why she had sent the object to
Ferreira in Montauk, New York, the older sister replied: "I'm not feeling
too good, and I don't have much time left."

Ferreira had eagerly shown the bright green object to neighbors, local
reporters, representatives from the local office of the Department of Natural Resources, and anyone else willing to take a look. Eventually, someone suggested to Ferreira that it might be ambergris.

"A hundred years ago, you would have no problem finding someone
who could identify this," James G. Mead, curator of marine mammals at
the Smithsonian Institution, told the *Times*. "More often, you have people
who think they've found it and they can retire, only to find out it's a big
hunk of floor wax."

Back on the beach in Brighton, under a ceiling of dark clouds, I have
made it to the end of the bay. I have nothing to show for the mile and a
half I walked. From the portentous clouds, I know a storm is on its way.
Hurrying now to make it back to my car, I still stop every few steps to
prod another random object with my driftwood. And as I do so, I wonder how to gain that experience for myself—the experience that so few
people have—to be able to identify ambergris.

A few days earlier, I had read another description of ambergris in *The
Emperor of Scent: A Story of Perfume, Obsession, and the Last Mystery of the
Senses*. Author Chandler Burr recounted a conversation with fragrance
chemist Luca Turin, who described a journey made by Guy Robert—the
French perfumer who, in the 1970s, created the fragrance Dioressence for
Dior—to London, to evaluate a large piece of ambergris.

Turin told him: "So Guy gets on a plane and flies up to see the dealer,
and they bring out the chunk of ambergris. It looks like black butter.
This chunk was about two feet square, thirty kilos or something. Huge. A
brick like that can power Chanel's ambergris needs for twenty years. This
chunk is worth half a million pounds."

In fact, established perfume houses like Chanel and Dior likely would have avoided purchasing ambergris that resembled black butter. A hundred years ago, the largest French perfume houses employed specialists with highly trained noses to evaluate and purchase ambergris directly from dealers. And they made sure to select and obtain only the very best ambergris. In the mid-1930s, two fragrance chemists named A. C. Stirling and William Poucher attempted to further classify ambergris, dividing it into ten distinct gradations, based on color, origin, and other characteristics. The differences between the several different grades of ambergris are narrow but deep: black ambergris is fresh and soft, and smells like sheep dung. It can be worked with the hands like wet clay and is soft enough that several pieces can be added to one another and rolled into pungent balls. The least refined grade of ambergris, it requires additional processing before it can be used to make perfumes. It lacks all the complexity of a small weathered piece of silver ambergris—the highest grade in Stirling and Poucher's classification system—that has spent decades in the ocean, slowly transforming one molecule at a time.

Back on Brighton Beach, I have finally reached the deserted parking lot and my waiting car. The wind is even stronger now, angling in sharply from the sea. The boughs of the slender pines growing near the grassy backshore are beginning to bounce with each gust. I am exhausted. I had prodded a few hundred random objects on the beach and pried up several heavy pieces of warped timber deposited on the camber of the shoreline by a recent storm. And I am more confused than ever: Am I looking for something that is luminescent and olive green, like the object Ferreira received from her sister in the winter of 2006? Or should I be searching for a substance that instead resembles a block of greasy black butter, like the ambergris in the anecdote Turin shared with Burr? Could it be both? Or is it, in fact, neither?

* * *

In March 2009, while swimming in the ocean near Oakura, near New Plymouth, on New Zealand's North Island, seven-year-old Ben Marsh swam straight into a large round lump of ambergris. A few days later, an article reporting the find appeared in the *Taranaki Daily News*, beneath the headline "BEN STRIKES IT LUCKY WITH SMELLY FIND."

Marsh had seen the ambergris floating in the waves, picked it up, and carried it from the ocean. The newspaper article was accompanied by a photograph of blond-haired Marsh, proudly cradling his ambergris: a potato-size lump that filled both of Marsh's small hands. It had a dark and shiny surface, a patina, and, at each end, its surface was white, as if it had polar icecaps. It weighed a little more than eleven ounces. Also visible in the photograph is the boy's mother, Julie, who told reporters that the lump smelled strongly, like manure, and would have stayed on the beach had her husband, Nigel, not listened recently to a radio show about the value of ambergris. The lump had been appraised by an ambergris trader, who confirmed its authenticity and, the Marsh family hoped, would soon be extending an offer of purchase. The potential value of the ambergris had come as something of a surprise to Ben Marsh.

"I thought," he told a reporter, "it was just an ordinary floating rock."

*   *   *

By the time I read about Ben Marsh's ambergris, I had begun a transformation of my own. I had been searching for ambergris for more than six months—since the moment I first read about the half-ton boulder of tallow that had washed ashore on Breaker Bay, near Wellington. Gradually, as the months passed, I had become tight-lipped and territorial about it too. It began slowly. At first, I started to eye other people on the beach with suspicion. After several months of walking the local beaches, this had become, in my mind at least, my territory. It was my patch. One day my wife and I saw three people on the beach at Aramoana, watching a scrappy little dog dig into the flat wet sand. The dog was almost out of sight, at the bottom of a rapidly growing hole. The sky was an unbroken blue, but it was cold and windy on the beach. The people had turned their backs to the surf and were instead facing the dunes and the grassy banks of the backshore, glancing at their watches and holding empty bags. Clearly, they were searching for *something*. Wet clumps of sand flew through the air. We approached them across the beach. I tried to adopt a friendly *what a beautiful day for a walk on the beach* tone of voice and asked them, "What are you looking for?" They said nothing. The surf broke behind us on the beach. A gull flew past in the wind. The people stared at us without smiling and finally answered, "Nothing."

Maybe they were looking for *tuatua*, a flavorful type of shellfish found on flat sandy beaches like this one. At other times, I had seen people combing the beach for shells or harvesting slippery armfuls of fresh seaweed from the sand and carrying them away in dripping bags. But per-

haps they were looking for ambergris along the tide line, right here on
my patch. It felt like a violation.

*   *   *

Eventually, several months after beginning my search for ambergris, I
contact Adrienne Beuse, a full-time ambergris dealer based in Dargaville,
about a hundred miles north of Auckland, and find that she is willing to
talk with me. Together with her husband, Frans, Adrienne Beuse runs
one of the most successful and high-profile ambergris trading companies in New Zealand, selling ambergris to buyers spread across the world.

"It's not the only incident John Vodanovich has been involved with,"
Beuse tells me, when I ask her about the battles that had unfolded on local
beaches in 2003. "But that's the most notorious one." Along with several
other local residents, Beuse claims, Vodanovich is part of a large, semi-
organized collective—what she calls a gang—that collects and trades ambergris aggressively from Northland beaches. "They're called the Beach
Mafia up here," she says matter-of-factly. "They claim a proprietary interest in the beach. They are defending, I guess in their minds, their territory. And it's worth a lot of money. If a piece worth $50,000 washes up,
they don't want anyone else to find it."

Beuse pauses, as she tries to find the right words to describe the situation in Dargaville, with so many different factions searching for the
ambergris that washes ashore there. "Trouble," she says eventually, "can
ensue." Sherman's injuries, she tells me, guffawing loudly as she does so,
were not serious: he had been left with a clearly visible set of tire tracks,
she says, imprinted across the backs of his legs.

A small city with fewer than five thousand residents, Dargaville is
mostly an uneventful place, situated on a slow, wide bend of the Wairoa
River, as it winds toward the coast eight miles westward. Nothing much
happens. Nevertheless, in late 2003 a quiet sort of war really was taking
place on its beaches, in front of the pounding surf. Most of the local residents were oblivious to it, but a current of threat ran along the shoreline. Posture was sufficient to win most battles, and owning a ferocious-
looking dog helped. But, inevitably, the threat sometimes exploded
into incident. The Beach Mafia occasionally had to act to retain control
of local beaches. If the weather conditions were right for ambergris—
sustained periods of strong westerly and southwesterly winds, which
cause high tides and rough seas—several separate groups of ambergris

collectors, each with their own allegiances, would make their way to the coastline to harvest ambergris. At those times, the potential for violence became palpable.

More than once, the local police had been forced to intervene. Adrienne and Frans Beuse had been threatened and then attacked while driving their all-terrain vehicle along the shoreline near their home. Two speeding cars had pursued them along the beach, edging closer and closer, almost forcing them into the ocean. The incident was reported in a September 2004 article in the *New Zealand Herald* under the headline "WHALE WASTE CAUSING BEACH DISPUTES." "When we reported it," Adrienne Beuse told the *Herald*, "we were told by the police to either stay away from the beach or to take several people with us if we wanted to go down there but we feel sooner or later there's going to be a tragedy out there."

Gripping the wheel of one of the cars that sped across the sand toward Beuse, she says, herding her toward the ocean, was John Vodanovich. "He's still around, and a major player in the industry," she tells me. "These northern beaches are, you know, premium collecting areas, and there's a bunch of people that rely on them. In fact, someone came to me the other day and there was some good weather, and they said some groups were running into each other on the beach. And they figured they would go and get a deck chair and set it up on the beach, because something good was going to happen."

*   *   *

Every morning I make an eight-mile journey to work, driving south from our tiny house in the historic port town of Port Chalmers to Dunedin, along the twisting harbor road. Sometimes I ride the bus to work instead, past cramped little houses squeezed into suburbs: Sawyers Bay, St. Leonards, Roseneath, Burkes, and Maia. Smoke curls from chimneys into a wet gray sky. Peeling wooden boats sit on trailers in driveways. A handwritten sign on a gate reads: "*Horse Poo $2 a bag.*" Overgrown clumps of flax line the pavement, their spindly black stalks bent over the roadside by the wind. Through a thick green screen of trees and bushes, I watch the water closely. To the north, churning past the heads, fresh tidewaters surge into the harbor, refilling the rocky bays. In the middle of the channel, the water is brisk and blue. Seagulls have settled on the waves, like grains of salt sprinkled on the dark water. On cold winter mornings, the lights of Broad Bay, Portobello, and Macandrew Bay across the harbor twinkle in muted yellow clusters through the fog, like unnamed constellations.

The entrance to Otago Harbor a few miles farther to the north is narrow and acts like a natural bottleneck. Immediately after the harbor opening, the channel widens briefly before narrowing again, as the water is forced between two rocky peninsulas, through a tideway almost completely obstructed by two islands—Quarantine Island and Goat Island—that sit green and round-shouldered on an anticline in the strait. In short, the harbor is a thirteen-mile-long bottle. Its contents are completely replaced every twenty-four hours. From above, it resembles a sea horse, clearly divided somewhere near its middle, after which its tail unfurls to the southwest and terminates finally at Dunedin.

Occasionally, things get trapped in bottles, able to enter but not exit again. Perhaps the tides have carried pieces of ambergris through the tight opening of the harbor, and they have become trapped somewhere in the natural green net strung across the channel. In fact, it isn't difficult to imagine countless pieces of ambergris—heavy, misshapen boulders of it—littering the undisturbed rocky shores of Quarantine Island in the middle of the channel.

I have read on the website of French ambergris trader Bernard Perrin that weathered, waxy lumps of ambergris floating in the ocean can be identified from the shore by a trained and careful eye, and will be surrounded by a ring of calmer water. The hydrophobic organic compounds in the ambergris—the alcohols, ketones, lactones, and aldehydes—disrupt the surface of the water so that, from a distance, it resembles a little localized oil slick. I scan the channel beyond the cabbage trees and the flax: seagulls bobbing on the waves, fishing boats, green rocky coastline. But this morning there are no unexpected patches of calm water amid the whitecaps. In other words, I will not have to explain my actions to my wife, who has asked me several times what I will do if I actually see an area of oddly calm water in the harbor as I drive to work in the morning.

*　*　*

One hundred twenty-five years before Ben Marsh collided with a lump of ambergris in the water off Oakura, someone in nearby New Plymouth devised one of the strangest ambergris-related business schemes that I have come across. The plan was reported in the *Wanganui Herald* in September 1883. "I hear they have an idea at Waitara of developing a new industry there," wrote the *Herald*'s correspondent. "A few young whales are to be captured and literated in the river—the bar preventing their return to the ocean—where they will be fed on a preparation of chloridine. This, in course of time, will produce vast quantities of ambergris, which commands almost a fabulous price in the home market."

But ambergris cannot be farmed like potatoes. It cannot be produced by intentionally poisoning sperm whales. It must be found. And finding ambergris is a little like finding gold. In the mid-nineteenth century, across the wide and empty plains of the American West, fortunes were made and squandered by those who turned the soil over and saw the dull gleam of gold glinting back at them through the dirt. But there are important differences between ambergris and gold. Unlike ambergris, gold and other mineral resources are predictable. Their formation, millions of years ago, required a specific set of geological conditions to occur all at the same time. Knowing the geological conditions that produced them increases the likelihood of finding deposits that can be mined and extracted. This could not be less true for ambergris. Its formation is mysterious and still debated among cetologists, and knowing the precise conditions under which it is formed helps very little. It is simply out there somewhere, riding the cold waves until the correct conditions combine with one another and bring it to shore.

The most important distinction is that, under almost all circumstances, mineral resources like gold stay precisely where they were made. Locked within the rock, they remain unchanged and unmoved for millions of years. The only challenge is to locate and extract them before someone else gets to them. In contrast, ambergris is always changing. The challenge is to locate it before it is eroded by the elements into pieces that are infinitesimally small and worthless. Eventually, it will become nothing and will simply disappear. And ambergris is always moving. It is highly mobile. It might be found tens of thousands of miles from where it was made, decades later. Imagine how different the history of the California Gold Rush might have been if the gleaming nuggets of gold that the prospectors were seeking traveled vast distances, were likely to move again, and were eroding as the hopeful prospectors searched for them.

Nevertheless, in January 2006, Leon and Loralee Wright must have felt like they had found gold. During a walk on Streaky Bay, in remote southwestern Australia, they stumbled over a boulder of ambergris weighing thirty-two pounds. It didn't gleam like gold. In fact, they told the Reuters news agency, it resembled a tree stump. It was odd-looking, they said, and not worth a second glance. They left it on the beach. As they walked away from it, a dizzying number of destinies waited for the object in Streaky

Bay. It might be picked up again twelve hours later by the following high tide. Afterward, it could reappear somewhere else, in another bay farther along the coast, or be borne away on the open seas to drift again for years. At sea, it could be caught in a net and hauled out of the water, or broken up into smaller pieces by rough seas. Or it might stay exactly where it is, left to erode on Streaky Bay, a remote and windswept place that sees few visitors.

Instead, the Wrights were lucky. Two weeks later, they returned to the beach. Between visits, they had carefully researched the pungent-smelling lump, and they now suspected it was ambergris. When they made their way back to the beach, the lump was still there, where they had left it. "We were sort of dancing and clapping and cheering on the beach," Loralee Wright told National Public Radio in January 2006. "We were very excited."

If they had waited for another two weeks, it might have no longer been there. Perhaps the shifting sands would have buried and hidden it from sight. This time they took no chances. Loralee Wright convinced her husband to load the strange object into their car and take it home. Describing the object in an interview with NPR, Leon Wright said it looked like "either a gray rock or a burnt-out stump. Very solid on the outside and very dry, but inside it's like a black sticky tar with squid beaks and stuff inside it. It actually smells like very sweet cow dung." Soon after hauling it away, the Wrights arranged to store the ambergris in the vault of a local bank, sealed in a plastic container. It was worth approximately $250,000.

"Lucky for us," said Wright, "the high tide didn't pick it up and the wind carry it back to sea."

* * *

A few weeks later, I found a photograph online of Loralee Wright, posing on the sand with her ambergris. She is crouched on Streaky Bay in southwestern Australia, behind a large misshapen black-and-gray boulder—an attractive, blond-haired woman, squinting into the sun with a forgotten pair of sunglasses perched on top of her head. The ambergris is so large that it almost hides one of her legs and casts a dark shadow across the lower half of her body. It is enormous.

"All these masses," Clarke had written in "The Origin of Ambergris" in 2006, "are thicker at one end than at the other. The thick end has a depression, a kind of concavity, except the thick end of the largest haul"—the largest piece of ambergris that Clarke personally inspected in

1953—"which is flat. The thinner end of each mass may taper, or be like a thick bobbin stuck out from the main mass, and its face is somewhat rounded or convex."

Wright's piece of ambergris is exactly as Clarke had described it: a dark-gray boulder with a thinner tapered end that points toward the cloudless blue sky like a stubby bobbin, and its flattened, thicker end, resting solidly in the sand.

When I contact Loralee Wright to learn more about the circumstances of her find, she replies: "Firstly, I really don't want to discuss the sale of it, but I can give you some information in regards to my piece of ambergris."

Back in January 2006, she says, after they had finished celebrating their find on Streaky Bay, the Wrights took their ambergris home and carefully stored it away. A month passed and the ambergris began to lose weight—just a few ounces at first, but then half a pound. A few months later, when it had lost more than a pound, the Wrights began to worry. They monitored it like nurses caring for a sick patient. In an attempt to prevent further weight loss, they swaddled the ambergris in cloth. But it continued: an ounce a week, every week. Gone. Evaporated. Ambergris is sold by weight, and lost weight means lost revenue. Finally, eight months after the Wrights found it on the tide line, their ambergris had lost almost two and a half pounds, or several thousand dollars in value. Every day longer that they held on to it, the ambergris became less valuable. Presumably, that was the moment they decided to sell it, before it lost more weight.

I ask Loralee if this is what had happened: no comment.

*   *   *

There was a reason for Loralee Wright's taciturnity: the legality, or otherwise, of importing and exporting ambergris between various countries. The regulations concerning international trade of ambergris are a tangled web of restrictions and treaties, with conditions and clauses that differ from country to country. It is almost impossible to find out what the regulations are, who decides them, which agencies enforce them, and how they propose to do so. In Australia, a particularly strict classification of sperm whales under the Convention on International Trade in Endangered Species (CITES) prohibits the importation and exportation of ambergris, which is why Loralee Wright had declined to comment on the sale of her ambergris. Commercial trade in ambergris is allowed under the CITES treaty in the United States, where sperm whales are classified slightly differently than in Australia, but prohibited by the Marine Mammals Protection Act of 1972.

This might sound like a simple-enough declaration, but it took several weeks to confirm even these regulations. In one Kafkaesque turn after another, I was referred from agency to institution, to agency. First, I contacted Charley Potter, the marine mammals collection manager for the National Museum of Natural History in Washington, D.C., who referred me to Monica Farris, a biologist at the U.S. Fish and Wildlife Service Division of Management Authority, who then referred me to Jennifer Skidmore in the Office of Protected Resources, part of the National Marine Fisheries Service. In every situation and in all locations, there were always exemptions, conditions, and loopholes to exploit. "In order to sell ambergris internationally from Australia," I was told by Michelle Scott, the assistant director of the CITES Management Authority of Australia, "the product would need to be classified as an eligible 'non-commercial' purpose. Under our legislation, the Environment Protection and Bio diversity Conservation Act 1999 (EPBCA Act), permits may be granted for specimens of a non-commercial nature."

If loopholes cannot be found, more extreme measures are taken. I had heard a rumor of a very large piece of ambergris that had made landfall several years ago in Australia, where exportation of ambergris is illegal. It was broken into manageable pieces, loaded into five bulging suitcases, and then taken to the Middle East by Arabian diplomats.

* * *

When cold weather and driving rain kept me away from the shoreline, I stayed at home and studied Robert Clarke's scientific papers on ambergris: "A Great Haul of Ambergris," from 1954, and "The Origin of Ambergris," from 2006. The first is a short communication, barely one page long, published in the well-respected scientific journal *Nature*. It details the harvesting of a boulder of ambergris weighing 926 pounds from a large bull sperm whale in the Antarctic, in December 1953, by the *Southern Harvester*. A photograph of the ambergris appears beneath the first paragraph: a large black boulder with squared-off ends, like a solid cement block, suspended above the deck of the ship on a block and tackle. By coincidence, it has the same blunted shape as the head of a sperm whale. Thick coils of rope are wrapped around it, forming a makeshift sling. In the background is the curved, sunlit prow of a ship. A man

Ambergris weighing 926 pounds, taken from a sperm whale in the Antarctic on December 21, 1953, on board the floating factory the *Southern Harvester.* Credit: Norbert Dentressangle.

Boulder of ambergris weighing 342 pounds, taken from a sperm whale on November 21, 1947, and inspected by Dr. Robert Clarke. Credit: Robert Clarke.

stands to one side, frozen in profile, eyes hidden in shadow, arms cautiously held out to steady the ambergris. It is enormous. In fact, it was the largest single piece of ambergris ever harvested. It dwarfed both the king of Tidore's ambergris in 1693, which weighed 182 pounds, and the boulder weighing 180 pounds that Louis Smith crawled through a whale to harvest, on the beach in Recherche in 1891. The more recent article is definitive. It is fifteen pages long and includes tabulated results of chemical analyses of ambergris samples and magnified photographs of the stratified interior core of a piece of ambergris. There are more photographs too: one taken in November 1947 of a misshapen boulder weighing 342 pounds, resting on the deck of a ship, surrounded by a ring of curious whalers.

Together, these two articles represent some of the most careful scientific writing on ambergris. In fact, they represent some of the *only* scientific writing on ambergris. My attempts to acquire information elsewhere had been mostly unsuccessful. I was left with only two scientific articles—two odd but appealing academic papers, written in Clarke's formal tone, with more than fifty years between them. They were like nothing else in the scientific literature. Nevertheless, they were all I had.

* * *

It is now almost sixty years since astonished whalers removed the enormous 926-pound boulder of ambergris from a single bull sperm whale in the Antarctic. The era of industrialized whaling has thankfully ended. Sold to the Japanese, the *Southern Harvester* was finally broken up for scrap in 1971. But Robert Clarke lives on. From the ancient coastal city of Pisco, Peru, near the abundant marine life of the Paracas National Reservation, Clarke continues his research and still occasionally publishes his findings in scientific journals.

Across the decades he has seen more, and significantly larger, pieces of ambergris than anyone else alive: on the deck of the factory ship the *Southern Harvester* (a 342-pound boulder, inspected in November 1947); at a whaling station on the island of San Miguel, in the Azores (a 42-pound lump, examined in June 1949); at another whaling station farther west, on the island of Faial, also in the Azores (a 9-pound piece, inspected in July 1949); and the 926-pound boulder in 1953. And so, finally, I nervously moved aside Clarke's two papers and wrote via e-mail to ask him for a brief interview. Later that day, I received the following response, which for some reason, I found full of accidental poetry:

Dear Mr. Kemp,

I am answering your last e-mail on behalf of my husband, Dr. Robert Clarke, who is at the moment ill in bed. So, he will not be able to talk to you on the telephone. He is ninety years old and has a more or less serious disease.

With our best wishes,
Mrs. Obla Clarke

4 IT LOOKED LIKE ROQUEFORT AND IT SMELLED LIKE LIMBURGER

I've always been poor. Now I can build a home and educate my children. ∗
RALPH KENYON, unemployed father of newborn twins, speaking to
Time magazine in Bolinas, California (1934)

*I admit this is a smelly business, but it is my feeling right now that if a ton of
ambergris came floating by my ship, I would just spit on it and sail on.* ∗
ROBERT E. DURKIN, captain of the *William Schirmer*, in 1946 after the
smelly mass his crew hauled from the ocean and brought to Baltimore
proved to be worthless

Early in the afternoon on March 2, 1934, radio operator Alf Harrodon
was taking a walk along Bolinas Beach, about thirty miles north of San
Francisco, when he found something in the sand. It was a large gray object with a mottled surface. It smelled, Harrodon later told *Time* magazine, like Limburger cheese. Although it was cumbersome and soft—
and weighed around sixty pounds—Harrodon picked it up and took it
home. A sample of the object was quickly dispatched for analysis. The
verdict, when it came back the next day, was what everyone in Bolinas
had started to cautiously hope for: 70 percent ambergris. At the market
prices for ambergris—then $28 an ounce—Harrodon's cheesy lump was
worth almost $27,000. For context: when Ford began selling its Model
40A motor car that same year, it sold for around $500. If Alf Harrodon's
lump was genuine ambergris, he'd be able to buy a fleet of more than fifty
of them. Thus began the Ambergris Rush of '34.

As word of Harrodon's find began to spread through the poverty-
stricken Bolinas community, people stopped what they were doing and
rushed through the potholed streets toward the breaking surf. Doors
slammed across town. Workplaces closed early. Food burned in the ov-
ens. Before long, the entire town—all 250 residents—was on the beach,
shouting to one another. Workmen ran along the shoreline still clutching their tools, and housewives stood together on the wet sand, tidewater
darkening their house slippers. Coast Guard boats patrolled the choppy
waters. Even the schoolchildren were there—the school board had called
an emergency meeting and announced an impromptu school holiday,

so that everyone could join the search party quickly beginning to gather on the beach. The arrival of the ambergris was reported by newspapers across the country. A March 3 headline in the *New York Times* read: "FIND 'AMBERGRIS' ON COAST; THREE CALIFORNIANS MAY GET $38,000 FOR DISCOVERY."

Times were hard in Bolinas. That night, the beach was lit with bonfires. In the distance, burning torches bounced about strangely in the darkness. Empty-handed residents searched through the night for their own ambergris. And many of them found some. Eventually, a total of three hundred pounds or so of the substance was found along the Bolinas coastline. An unemployed man named Louis Pepper made his way to the beach, leaving his foreclosed home with his nine children trailing behind him, and found fifty-five pounds of it scattered on the sand. Local mailman Harold Henry found another piece of it and took it home to his pregnant wife. "Ronald Gandee, 24 years of age," reported the *Los Angeles Times* on March 8, 1934, "plans to marry Frances Longley, 20, if the seventeen-pound lump he found is genuine."

Elsewhere, in San Francisco, chemist Emory Evans Smith was in a tight spot that was quickly getting tighter. As the manager of the laboratory that first analyzed Alf Harrodon's object, Smith's verdict had sent 250 desperately poor people onto the beach, propelled there by the hope that they might be able to escape Bolinas. A few days before, the community had been so poor it was unable to raise $15 to pay for the chemical analysis of the cheesy lumps that had washed up on its beach. The *San Francisco Examiner* had paid the fee instead. Since releasing the initial confirmation of ambergris, Smith had begun to get more nervous about the pronouncement. And Bolinas residents had been arriving with other items for him to test: pieces of soap, potatoes, sponges, sewage, and a dead rat. "There is potential tragedy in this situation," he warned *Time* magazine.

Things were about to get worse. The Ambergris Rush began to spread farther along the Marin County coastline. And people were following it. More objects were washing ashore at Point Reyes to the north and also farther south, on Stinson Beach. "That stuff people are finding is not ambergris," Stinson Beach mayor Newmon L. Fitzhenry told the *Berkeley Daily Gazette*. "It is, in fact, the crust that forms inside sewers and is dislodged in cleaning. We know, because we had some analyzed several years ago." In total, the lumps found at Point Reyes weighed almost five hundred pounds. News of their arrival brought a steady stream of hopeful ambergris hunters to the coast: "Seven hundred persons were reported sifting the sands at Fort Baker," reported the *San Jose News*, on March 9,

1934. "Other hundreds invaded Bolinas Bay and Drake's Bay estuary, near Point Reyes, where the latest discovery was made."

Under the headline "AMBERGRIS HUNTERS HAVE HIGH HOPES OF WEALTH" in the March 8 issue of the *Berkeley Daily Gazette* is a photograph taken on Bolinas Beach. In it, teacher Loa Forsythe is surrounded by four schoolchildren on the beach, each holding aloft a white rounded lump of what they think is ambergris. In the background, people are bending over and elbows are pointed skyward in the search for ambergris. The sun is bouncing off the excited cheeks of the children. They believe they are leaving Bolinas for a better life. Representatives of the Bouquet Perfume Company provided encouragement when they arrived in Bolinas willing to pay local residents the market price for samples of the substance.

The following week, it was all over: a case of mistaken identity. The residents of Bolinas were, in fact, the proud owners of approximately three hundred pounds of chemicals that had been used to clean San Francisco's sewers. Mayor Fitzhenry had been right. For a week, the iceboxes of Marin County had been used to store cheesy lumps of encrusted sewer cleaner. Evidently, it had been discharged into the ocean around San Francisco, and currents had then carried it north, depositing it at various locations along the sweeping coastline: at Stinson Beach, in Bolinas Bay, and farther north at Point Reyes. People were angry. For a moment, they had been given hope, and then, just as quickly, it had been stripped away again. In desperation, they refused to believe the new findings. From the March 13 issue of the *Berkeley Daily Gazette*: "The 'ambergris rush,' at its height akin to the gold rush of 1849, brought hundreds of hopeful searchers to the beaches to pick up pieces of the stuff which looked like Roquefort and smelled like Limburger. Health officials warned of the danger of contagion in the substance the searchers found but the warnings were disregarded."

The people of Bolinas were poor and desperate. They had been fooled by sewer cleaner. And in Bolinas they all would stay.

*   *   *

The short-lived elation that spread through the streets of Bolinas in 1934 is an integral part of searching for ambergris. By now, I had experienced it several times myself. It begins with a quick survey of the tide line, perhaps too quick, followed by a careless misidentification of an object—a piece of sun-bleached driftwood or a rounded pebble of chalkstone—

that has come to rest on the gentle camber of the beach. It looks like ambergris. It is a heart-stopping moment. I bend quickly to retrieve it, hurrying to hold it up to the coastal light for closer inspection. Realizing my error, the familiar disappointment returns and, along with it, the tidal chill to the bones.

Moments like these are inevitable. They happen every week. As often as possible, I consult the tide tables and arrive on the coast just after high tide has peaked, when my chances of finding ambergris are the greatest. It might sound like reductive logic, but to find something like ambergris, which washes ashore with the tide, one must simply spend a lot of time on the beach. The search is no more or less sophisticated than that. But the moment I finally realize this, standing on Long Beach late one afternoon, waiting for another golden wave, is filled with Zen. I grow to appreciate the moments *between* waves—savoring the handful of half-silent seconds, when the possibility of still finding ambergris outweighs the disappointment of not yet finding it.

During the early months, I must have logged several hundred miles, trudging along the coastline in sometimes difficult conditions. Searching for ambergris became an absurd but irresistible impulse. Every moment that I was not out there, I felt certain that I was missing the one perfect wave—an otherwise unremarkable surge in the tides that would deliver an enormous boulder of ambergris onto the sand. If I was not there, I would miss it. At home, I sorted through flotsam: a seemingly endless supply of burnt wood, shards of bone, beach glass, and little fragments of plastic. Occasionally, I kept a few unidentified items—especially those that most resembled the strange, grainy photographs of ambergris I had found in scientific journals and textbooks.

One day on Tomahawk Beach, as dark rain clouds unfurled across the horizon like dirty linen, I found a dead seal pup, half-buried in the sand. I walked on, following the sweep of the bay. Farther along the beach I stopped to inspect pinecones, a glove, and an old coconut husk. It began to rain again. Farther still along the beach, with the tide quickly receding, I found a dead stiff-legged starfish, the color of a gun barrel. I dropped it into a plastic bag that was tied around my wrist. Zigzagging my way steadily northwest across the wet sand, I arrived at the end of the beach, bending to pick up a dinner plate-size bull kelp holdfast, which I threw into the breakers like a dented and buckled discus.

Over time, the objects I took home gradually changed. It was a period of refinement. I began to recognize each type of shell and seaweed, gradually imposing taxonomic order on an unfamiliar environment. I learned to recognize the common bird species, pointing them out to my son as he slept in his stroller: Arctic terns, cormorants, oystercatchers, spoonbills, and white-faced herons. Once, we saw a yellow-eyed penguin, rushing along the beach into the late afternoon sun like a round-shouldered commuter. We became hopeful beachcombers. I knew the difference between an oyster shell and a mussel shell, and could tell the flat rubbery blades of bull kelp from the thin wrinkled fronds of bladder kelp. One afternoon, finding a delicate little basket of bone in the sand, I was able to identify the skull of a black-backed gull, picked clean by the crabs. And from its design, I could differentiate between a recently discarded Coke can and a vintage, well-traveled can that had spent months in the ocean. But I found no ambergris. All those months and the steady miles I spent leaning into the rain, and I was still empty-handed.

*   *   *

Once one has acknowledged its innumerable mysteries, ambergris is just another piece of flotsam, transported by the ponderous and unpredictable movements of the ocean. It arrives on the shore via the same circuitous route taken by the multitudinous other objects I have found there: the lone shoes, the half-eaten fish heads, the bent and rusted umbrella frames stripped of their canvas. But it is the rarest and most valuable flotsam. If a boulder of ambergris weighing two hundred pounds washes ashore on Long Beach, it could be worth, depending on its quality, at least $2 million.

Many of the objects I find on the beach have spent a considerable time—sometimes years and even decades—trapped in huge swirling ocean currents called gyres. The existence of the gyres has become more well-known in recent years, since the discovery of the Great Pacific Garbage Patch—a tangled, Texas-size accumulation of marine trash in the central North Pacific Ocean. There are, at different locations in the world's oceans, five major oceanic gyres: rotating vortices that measure thousands of miles across.

A gyre functions, oceanographer Curtis Ebbesmeyer tells me, like "a giant clock. A turbine. It's a giant gearbox."

And Ebbesmeyer should know: he's a self-proclaimed flotsametrician. For more than forty years, the coauthor of *Flotsametrics and the Floating World: How One Man's Obsession with Runaway Sneakers and Rubber Ducks Revolutionized Ocean Science* has studied how flotsam drifts in the world's oceans. Despite his efforts, he says, almost nothing is known about how long an object can float at sea or the path it will take while adrift. "Nobody really knows," he admits. "There are all kinds of scenarios."

Every object behaves differently at sea. On certain beaches, shoes for left feet appear ten times more often than shoes for right feet. Left-handed gloves drift in one direction; right-handed gloves in another. Trees float for about ten years before becoming waterlogged and sinking, or breaking apart in the water. Bottles last longer. After several decades, a sufficient number of barnacles have anchored themselves to the bottle and it begins to sink. Eventually, its journey comes to an end on the seafloor. Other drifters—as Ebbesmeyer calls them—simply make landfall and eventually disintegrate under solar radiation.

"We've had messages in bottles come back after eighty years," says Ebbesmeyer. "We know that about twenty species of tropical seeds can float for at least thirty-nine years by tank tests. My friends and I have been running tests for thirty-nine years. They're still floating."

Eighty years at sea: a human life span spent adrift in the ocean. For simplicity, let's assume that ambergris floats on the ocean surface for a decade. "Ten years is enough time to drift halfway around the planet," explains Ebbesmeyer. "The great gyres in the Pacific take six years to go around, so your ten years is about enough time to go twice around those great Pacific gyres, or three times around the Atlantic gyre, or around the five gyres that have three-year periods."

Once ambergris, or any other piece of flotsam, enters a gyre, it travels on average between five and eight miles a day. In one year, a slowly maturing boulder of ambergris might travel twenty-five hundred miles. But it is a journey about which flotsametricians like Ebbesmeyer know almost nothing. He can never know, for instance, precisely where in the vastness of the ocean a lump of ambergris began its journey, or where it has traveled to since. Even if he knows where a large piece of ambergris washed ashore, Ebbesmeyer would be no closer to understanding where it had come from or how long it had spent in the ocean before it arrived. He would simply know one important but infinitesimally small data point in its oceanic journey: where it ended. The rest of the journey is a mystery, an oceanic enigma. There are endless variations. It takes an object fourteen years to orbit the Melville gyre—a large elliptical current that sits in the Arctic Ocean, north of Siberia—but, once trapped in it, a drifter travels less than a mile a day. An object in the cigar-shaped Turtle gyre, which extends from the West Coast of the United States all the way to South Asia, travels six miles a day; glass balls have been recovered after orbiting the Turtle gyre nine times—a journey that lasted almost sixty years.

"To me," says Ebbesmeyer, "the ocean is kind of half random and half deterministic."

By concentrating on the more deterministic aspects of ocean currents, Ebbesmeyer and Jim Ingraham, a National Oceanic and Atmospheric Administration oceanographer, have generated a computer-based modeling system to track the movement of flotsam, called the Ocean Surface Current Simulator, or OSCURS. "It's a massive database," Ebbesmeyer explains. "And to get all that data in there, and clean, and then find the algorithms that represent the wind and the currents is a major undertaking. We were never funded to do that, so we just did it on our spare time."

I wonder aloud to Ebbesmeyer whether OSCURS would be able to predict the best places to find ambergris. "It's very crude," he tells me. "OSCURS is just a model for the North Pacific, and there really is no OSCURS for any other gyre on the planet. It took Jim Ingraham his whole life, forty years of work, to get that far."

Perhaps, like a message in a bottle, ambergris drifts for longer, even eighty years or more. "There are very few drifters that can last that long," says Ebbesmeyer. "These are like planetary drifters. They go a long, long, long way. It's a message to the future. It's almost like some of the satellites we put out to the stars. They're still out there. Some of the satellites are still transmitting after thirty years. Tropical seeds can float for at least thirty-nine years, and it looks like ambergris is in that very, very rare category of drifters that can survive for as long as a human might live, or as long as a whale might live."

Ebbesmeyer pauses for a moment. We are both, I think, considering the immense distances that ambergris might travel after spending so many years at sea. A hundred thousand miles in forty years. In eighty years, a distance of two hundred thousand miles—almost the distance from the Earth to the moon.

"Suppose we were both whales," Ebbesmeyer says finally, "and we were swimming around and we let some ambergris go. That would be our epitaph. They're like little floating tombstones of whales."

*　*　*

Located on the east coast of Otago Peninsula, Sandfly Bay is a long wide belt of sand that sits between two steep headlands, book-ended by rugged cliffs and enclosed along its curving length by a natural sandy wall. The shoreline cannot be reached by road. I walk along a gently sloping gravel path with my wife and son, dropping through grassy meadows filled

with grazing sheep. We trudge down a precipitously steep and sandy in-
cline, best descended by taking several patient, slow lunging steps, my
son giggling in a backpack behind me. I lunge, sink into the soft sand,
and then lunge again, until I stand at the bottom of the slope with sand-
filled shoes.

Out to sea, waves crash against a jagged remnant of volcanic rock in
the hazy distance, throwing a skirt of white water around its black base. It
is a beautiful and unspoiled place. Strong crosswinds whip the water into
a rough chop. A solitary wispy cloud unfurls high above us in the blue
sky. I watch as it drifts over the grassy dunes and toward the cliffs. It is, I
think to myself, almost whale-shaped: a good omen.

At night I have begun to dream of ambergris, of stumbling over a large
gray tortoise-shaped lump of it hidden in the dunes on Long Beach, or
a few miles to the southeast at windswept Aramoana, or on the other
side of the peninsula on Allans Beach. In the middle of the night, I lie
in bed and imagine, just a few miles to the north, out beyond the harbor
in the open sea, a slow-moving pod of sperm whales swimming through
the black water, past the darkened hillsides. University of Otago professor Steve Dawson had told me that sperm whales almost certainly pass
through Otago waters from time to time, headed farther north to Kaikoura—a whale-watching destination popular with tourists—or south,
to colder waters. No one really knows. Without trailing a hydrophone
from a boat into deep water to listen for their distinctive volleys of vocal-
ized clicks, it is impossible to know.

We search the southern end first of the bay first: a dead cormorant,
with dirty matted feathers; a flip-flop; dried green-black streamers of
kelp; a cassette, with its tape missing from its spools; pale gray clumps of
marine sponge; and empty mussel shells. I find a dead sea horse on the
beach. It is six inches long and dried and stiff. Sand has collected in the
depressions of its corrugated body. The end of its tail curls into a delicate spiral. Its gray eyes are sunken in their sockets. It smells fishy, but I
keep it anyway, carefully folding it into the deep pocket of my raincoat.
In more than six months spent searching for ambergris, it is the closest
thing to treasure that I have found. I leave my wife and son in the shelter
of the grassy wall and begin to head north, crisscrossing the sand, trying
to cover as much ground as possible as the tide recedes. The only sound
I can hear is the unbroken roar of the sea.

Halfway along the beach, at the edge of the wet sand, I see three or
four pieces of something that looks like driftwood—irregular, flattened,
almost circular in shape, and gray-brown against the wet sand. Bending

over, I take a closer look at one of them. Perhaps, I wonder to myself, it is fresh ambergris that has been dumped on the shore by the last tide. In a few seconds, I have eliminated all the substances usually mistaken for ambergris: it isn't a white, greasy piece of tallow, a dead gull, or a marine sponge. It isn't driftwood or a long rotting stalk of bull kelp. I feel my heart begin to quicken at the possibility that, after all these months, I have finally found ambergris.

Leaning closer, I extend an exploratory finger to give it a tentative prod. Unexpectedly, my fingertip enters the lump and then passes through it, leaving a long green painterly smear through its middle. I straighten slowly. In the middle distance, three huge sea lions lie on the beach. They huff and belch and steam in the sun. A large male raises a black flipper in the air and yawns, disturbing a cloud of flies. There is, closer to where I am standing, a long ladder of peaks and troughs in the sand—the distinctive path left by a sea lion as it lumbers back into the sea. With mounting horror, I look again at my finger: a dark thin green seam now trapped beneath my fingernail. And in the distance, the belching sea lions. Slowly, I begin to realize that I have, with the tip of my finger, sampled a fresh wet lump of sea lion shit. A long rolling breaker collapses on itself in the sun. I run to the water with my finger held aloft. A lone seagull watches impassively as I hold my hand in the ice-cold surf.

When my finger is clean again, I walk back along the beach empty-handed, past the dead cormorant and the driftwood, to my waiting family. It is late in the afternoon. Back at the southern end of the beach, shadows slowly slide down the cliff face as the sun makes its way toward the sea. In the pale blue sky above me, the whale-shaped cloud is gone.

*   *   *

In 1729, when Caspar Neumann wrote his monograph in *Philosophical Transactions*, the true source of ambergris still remained a mystery:

> There are few substances concerning the origin of which there have been so many various Opinions among authors. One ascribes it to the Vegetable Kingdom, another to the Animal, and a third to the Mineral Kingdom. But others not contented with the 3 Kingdoms, into which all natural Bodies are commonly reduced, have thought fit to make it a Subject of an Aereal Kingdom; and others again will have it to belong to none of these Kingdoms, but to a Marine Kingdom: and yet the whole Sea, with all its various Contents of Animals, Fishes, Shells, Plants, Stones, Waters, Salts, etc., may be ascribed to

one or other of the three usual Kingdoms; and therefore there is no Need of any such new Distribution.

Neumann was an exhaustive compiler. He managed to compile a list of almost twenty possible sources of ambergris, all of which had been proposed by various scientists during the previous century or so. They ranged from the ridiculous to the absurd. It was, some claimed, the sperm of a whale. Others believed it was a grounded meteor or a certain type of mushroom, which grew at the bottom of the ocean, floating to the surface after stormy weather. It was the fruit of a tree that only grew on the coast in Guyana or the liver of a particular species of fish.

Other writers added to the theories. In 1764 the anonymous author of *The History of the Discovery and Conquest of the Canary Islands* described a small sandy bay on the island of Graciosa, which was known by the natives as Playa del Ambar. "Here is sometimes found a very good kind of ambergrease," the author wrote, "in form something like a pear, having commonly a short stalk: by this is should seem that it grows on the rocks under-water, which are near to this place, and is washed ashore by the waves, for it is generally found after stormy weather."

Jean-Baptiste Denis, personal physician to King Louis XIV of France, proposed that ambergris was a mixture of wax and honey that had melted in the sun, fallen into the sea, and been transformed by the seawater and the action of the waves. "This opinion seems to be further supported," states an entry in *The New Universal English Dictionary* from 1760, "in that large pieces have been found before it has arrived at its full maturity, which being broke had wax and honey in the middle of them." Writing in *The History of Japan* in 1690, the German naturalist Engelbert Kaempfer concluded that ambergris was "a kind of bitumen generated in the bowels of the earth, or a subterraneous fat, grown to the consistence of a Bitumen, which is by subterraneous canals carried into the sea, and there undergoes a farther digestion, being by the admixtion of its saline particles, and the heat of the sun, changed into Ambergrease."

Others claimed that ambergris was, as Neumann indignantly put it, "the Dung of Birds." He continued:

Nay, they go so far as to describe the very Bird from which it proceeds. They say it is the Size of a Goose, with beautiful Feathers, and Spots, and is called in the Maldivian tongue *Anacangrispasqui*, and, in that of Madagascar, *Aschibobuck*. *Ferdinand Lopez de Castagneda* and others affirm, that this Bird feeds upon various fragrant Herbs, and that it deposits the precious Dung proceeding from them, on Rocks and Stones in and about the Sea.

In fact, when Portuguese explorer Duarte Barbosa encountered Maldivian natives in 1518, that was what they told him. The truth was perhaps stranger still, but, at least for the meantime, it would remain a mystery.

*   *   *

One wet Sunday morning, instead of driving to the coast, I call John Vodanovich, the Dargaville-based ambergris hunter. Several years earlier, he had jousted—vehicle versus fishing equipment—with Ross Sherman on Baylys Beach. Back in 2003, Vodanovich had been a professional ambergris collector. I want to know if he still searches for ambergris on Northland beaches.

It is midmorning. Vodanovich sounds either very tired, slightly drunk, or both. But he is willing to talk. Each question I ask him is followed by a long static-filled silence. I hold the phone to my ear, waiting patiently for a response. Birds twitter outside in the trees. Somewhere on the street, a car horn honks, waits, and honks again. And then Vodanovich slowly begins to answer a question, groaning at first, as he marshals the words. We speak this way for almost an hour. It is exhausting. And it's difficult to believe he is an important member of a dangerous and territorial Beach Mafia.

Collecting ambergris, Vodanovich tells me, is still his full-time profession. "I got threatened at Taharoa once," he says, "and got kicked off that beach. These Maoris followed us. They came and they just told us we'd got ten minutes or half an hour or something to get off the beach. We'd already done it though. But that was a bit scary. They must have got wind we were looking for ambergris, and they come down really quick, even though, you know, it's such a small beach there wouldn't be much there at all, I'd say."

Winter is the busiest time for professional ambergris collectors. Persistent westerlies dump an astonishing array of flotsam onto the shore, and hidden among it are lumps of ambergris. If the weather conditions are right, Vodanovich will spend a lot of time on the beach, walking the shoreline after each high tide to scour the sand for ambergris.

"Up here, I was one of the first that really got looking for it, hard out like that," he says. "When I used to go to, say, Ninety-Mile Beach, there was no one looking. Now you go up there and everyone is looking. Ten years ago, I had probably half a dozen people working with me. They basically all shat on me, everyone that's done it. They all just go out and do it on their own once I've shown them how to do it. Most of them didn't even know what it looked like, you know? My brother buys it off us now and sells it for us. Every day he gets odd-looking things, pictures sent to him, you know. He used to get bits sent to him that were horrible-looking bits of stuff, you know, and ninety-nine percent of it isn't ambergris."

A week earlier, the weather conditions in Northland had begun to change. The wind turned and started to come out of the west, across open water. Heavy seas pounded the coastline. The tide levels were particularly high. Before too long, Vodanovich began to enjoy the benefits, finding a piece of ambergris on the beach near his home that was, he says, "probably about the size of a tennis ball, or a bit more. It's probably worth about a thousand dollars."

And then, without warning, the winds changed direction again. The seas were calmer. I ask Vodanovich if he will be on the beach again later that day, to sort through the kelp and other flotsam on the shore.

"No," he replies. "The winds are wrong. And we've sort of milked this beach pretty good."

* * *

John Vodanovich's knowledge of the local beaches, and of the tidal and seasonal changes that take place on them, meant he knew precisely when to look for ambergris and exactly where to find it. In short, he had the sort of experiential knowledge that I would only gain by searching for ambergris for decades. Not everyone is as fortunate, or as experienced, as Vodanovich.

Anton van Helden, the marine mammals collection manager for the Museum of New Zealand, a modernistic six-story building on Wellington's harbor, is frequently contacted by beachcombers who hope he can identify objects they have found on the shoreline, which they believe are ambergris.

"People bring in a huge array of different things," he explains. "One of the most common things is glass sponges, but other things: lumps of tallow, lumps of microcrystalline wax that have come from industry." It happens several times a year, especially after heavy storms have passed across the Kapiti coastline, resulting in rough seas and higher tides than usual. People who think they have found a piece of valuable ambergris are, in most cases, carefully cradling dog feces, coal, vegetables, rotting seagull, soap, industrial waste, old whale blubber, eroded rubber, seaweed, pieces of tallow, or part of a greasy, decomposed sheep carcass. Filled with expectation, they wrap the object in tissue and store it in the dark, where

they think it will be protected from further degradation and loss of potential value. Once in a while, perhaps, they unwrap it and show it reverently, nestled in their cupped hands, to their friends. They whisper
around it. They weigh it. They smell it. And then, finally, they bring it to
van Helden so that he can assess it for them.

On average, he gets five such calls each year. But none of the objects
brought in for appraisal has actually been ambergris. "In the twenty
years that I've been here and people have been coming in with lumps of
stuff that they've found on the beach presuming it to be ambergris, not a
single piece has been," van Helden says. "Not a single piece. And it's interesting because I look at other people around the world who do similar things, and they've had exactly the same response, you know, in the
time that they've been doing it—not a single piece found on a beach has
been ambergris."

*   *   *

In March 2006, on a remote beach near Criccieth, in North Wales, dog
walkers Sean Kane and Ian Foster found two large mysterious lumps,
which together weighed 110 pounds. Their discovery was reported in
the local *Daily Post* newspaper on March 4, 2006, under the headline
"WALKERS HOPE AMBERGRIS WILL NET THEM A FORTUNE" In a small
photograph that accompanies the article, the objects look like large misshapen bales of hay. Two days later another article appeared, with a less
hopeful headline: "BEACHCOMBERS TOLD DISCOVERY IS NOT VALU-
ABLE AMBERGRIS." Analysis carried out by chemists at Bangor University School of Chemistry indicated that the objects were man-made. They
were two partly eroded boulders of paraffin wax. After originating from
somewhere in the Atlantic Ocean, most likely a container ship, they had
been brought to the windswept North Wales coastline by Gulf Steam
currents.

"The common way of trying whether Ambergris is genuine or not,"
wrote Caspar Neumann in 1729, "is to run a hot Needle into it, when
something like melted Resin ought to stick to the Needle; or to throw
it upon burning Coals, or to melt a little bit of it in a Silver Spoon over
a Candle." This method of evaluating suspected ambergris is known,
by those individuals who move in ambergris-related circles, as the hot
needle test. Ambergris vendors like Adrienne and Frans Beuse still in-
struct potential sellers—people who think they have found ambergris—
to perform it on the material they have found on the beach. Genuine

ambergris will melt into a thick, oily fluid the color of dark chocolate, releasing a complex spectrum of odors trapped by the slow years it spent maturing at sea. More complex chemical analyses can be used to identify ambergris, but if an object fails this initial test, it is not ambergris. It is something else instead: a piece of kelp stalk or the by-product of an industrial process that has spent decades in the ocean, slowly becoming unrecognizable.

"This Proof indeed has its use," wrote Neumann, "but if you are not exactly acquainted with the Smell, and observe many other Circumstances, but only attend to this melting, you may be deceived; for the factitious Ambergris may answer this Trial." Many other substances, in other words, also melt on contact with a hot needle into a thick, oily fluid the color of dark chocolate. "The most convincing bit I had," van Helden tells me, "was a bit of dark rubber that had obviously been washing around and had become quite eroded. And it burned, you know? And it liquefied, yeah. And then you go: 'Oh, but it smells like petroleum.'"

Once an object has passed the hot needle test, it is time for a thorough appraisal. An ambergris vendor might take a small sample of the material and roll it between her thumb and index finger, hoping to feel the tackiness that the waxy outer covering of ambergris should impart on the skin. Here, experience is everything. Decomposing tallow almost certainly will impart a tackiness to the skin. And so will a dead fish. The vendor might ask: Does it float? Are there inclusions of squid beaks and clearly stratified layers? She will measure and weigh it, cradling it like a midwife handling a newborn baby. Each piece of ambergris is unique, and the specific characteristics of each sample must be evaluated. These properties reflect its particular journey, its years in the ocean. She might spend a while with it, thoughtfully smelling its dark pitted surface, savoring its specific odor profile in the same way that a viticulturist might savor wine.

* * *

Ewan Fordyce's office is located in a creaking Victorian-era academic building on the University of Otago campus, just down the hall from a small and cluttered geology museum. Inside the museum, rows of tall wooden cabinets are filled with pieces of petrified wood and dark little knuckles of shiny fossilized shell.

Arriving early for a prearranged meeting with Fordyce, I walk between the museum aisles, bending to read the handwritten collection notes

carefully placed beside each specimen. Outside, the sun is shining. It is a
bright and windy day. In one of the rooms farther down the hall, someone is playing what sounds like the soundtrack to a TV show or a Broad-
way musical. Snippets of dialogue and sound effects boom and echo dis-
cordantly through the empty corridors. But in here, it's as peaceful as
a chapel. The display cabinets stand silently, like neglected reliquaries.
The walls are decorated with maps: a large-scale geological map of Dunedin and the peninsula, divided and demarcated into dozens of shaded
wedges and segments, each representing a different rock type; next to it
hangs a brightly colored bathymetric map of New Zealand and the vast
unbroken southern seas around it.

In the corridor outside the museum, a glass-topped display case
stretches along the wall to Fordyce's office door and contains the fossilized fragments of a prehistoric species of giant penguin. Fordyce is talking with a couple of graduate students. They laugh. I wait and fidget by
the giant prehistoric penguin. The sea-worn objects in my bag grate au-
dibly against one another. I slowly begin to suspect I am a fool, holding
a bagful of rocks.

During the previous months, I had continued to pick up and take
home various objects gathered from the shoreline. To do otherwise
seemed like a failure of the spirit. A little gray cairn of them sat on a cabinet in my living room. Another neat row of them was lined up along the
windowsill above my kitchen sink. Others weighed down the pockets of
my jacket. If I drove too fast around a bend, more of them rolled noisily
around the trunk of my car.

Once, while crossing a busy street in downtown Dunedin, I had been
unable to resist picking up a smooth pale stone from the gutter, and then
carefully smelling it, while cautious shoppers watched me and placed
themselves between me and their children. All the evidence indicated
that I was looking for something that resembled, exactly and in almost
all ways, a rock. In some instances, ambergris looked like smooth gray
egg-shaped stones. In other cases, it was dark, jagged, and misshapen, like
a broken piece of volcanic rock, with the curved points of squid beaks
clearly visible on its surface. On the beaches, I had been collecting all the
objects I could find that most resembled the images of ambergris I had
seen. Now I stood in the corridor, holding a plastic bag filled with them.

A few minutes later, in the peaceful twilight of his book-lined office,
Fordyce is patient and soft-spoken. He sits impassively at his desk as I
unload my plastic bag onto his desk. Sand scatters across the desk like
spilled salt. Gray-bearded and in his fifties, Fordyce is an associate pro-

fessor and head of the University of Otago Geology Department and a research associate of the National Museum of Natural History at the Smithsonian Institution in Washington, D.C.

"Seven specimens," he says, organizing them into two unequal rows in front of him. A graduate student knocks and enters, then leaves again. She might have concluded that we were constructing a Zen garden together, so intently were we contemplating the irregular gray shapes arranged before us. "Number one," Fordyce says to himself, picking up the first object and weighing it in his hand like someone preparing to roll a die. He pauses to gather his thoughts. "It actually feels relatively heavy," he notes, "and if you look at it, you can see little glinting crystal faces. Just straight off, that's a sure sign you've got a rock, and it's probably a volcanic rock, with the crystals having formed from a melt."

He gently places it back on his desk. "So, originally melted rock," he concludes. He picks up the object next to it and brings it close to his face and squints at it, as if he has mistaken it momentarily for a telescope. It is the same as the first object, he says quickly, before picking up the next in line. "Number three, for sure, is a piece of mudstone with streaks of sand-size green mineral." He holds it toward me and asks, "See the little streaks running through here?" I nod. "The mudstone is originally a marine mudstone," he says, "probably very ancient, and the green mineral is a type of seafloor precipitate."

He replaces it and picks up another, identifying it with a cursory glance at its pitted surface: another piece of local volcanic rock. Moving quickly now, he picks up another. "Your number five is a very typical piece of basalt from the Dunedin region," he says, blinking at me from behind his wire-framed eyeglasses. "Dark-colored," he continues, running a stubby, thick-knuckled finger over its surface, "lots of iron minerals in it." He places it back on his desk with a clatter, spilling more sand, loosening the last remaining grains from the pores that pit its rough surface.

Picking up the next object, Fordyce turns it over in his hand. More than any other piece I had brought with me, it resembles the ambergris in photographs I have seen during my research. It looks like a small, worn-down, and rounded piece of chalk. "Number six is encrusted with precipitate of carbonate, lime, formed by marine organisms," he says, "just in the same way that we can see little bits of lime on some of these other chunks. I think it almost looks like it's sitting on a very weathered piece of volcanic rock. I just saw a little glint of a crystal there—crystals like sugar-size."

I am deflated. Another piece of volcanic rock. And far too heavy for ambergris. "Number seven is a piece of schist," Fordyce says, picking up the final object in front of him. "We can see it's actually got two layers in it, and this darker, resistant layer is very typical of schist, which tends to form stacks of alternating light and dark mineral colors. This would have come from Central Otago way, down the Taieri River, just washed out on the beach."

He reclines in his chair and makes a pink tent of his fingers. We sit together for a moment longer, the two orderly rows of rocks in front of us. "Numbers one, two, four, and five are from local volcanic sequences," Fordyce tells me. "Number three is ancient sedimentary rock, and number six I'm not sure about." With slightly less care than I had unpacked them fifteen minutes earlier, I drop each rock back into my plastic bag. Mudstone follows volcanic rock and schist, becoming a multicolored pile of rubble at the bottom. I think of the little stone ziggurats that sit on my windowsill at home, and the makeshift levee of pebbles that meanders across the top of my bookshelf. Outside, the sun is still shining, bright and buttery on the flagstones. A few miles to the east, the tide will be returning to the shore, rearranging the weathered driftwood into new shapes. Gulls will be patrolling the sand for anything edible. I should be out there too, overturning the flotsam on the shoulder of the beach and peering underneath it for ambergris, while sand fleas squirm and flip in the sudden sunlight.

I slowly swing a bagful of ordinary rocks from my arm, feeling its weight. I thank Fordyce, and I wonder whether, on my bookshelf at home, gathering dust among the little pebbles of schist and the green-streaked mudstone, a fossilized and gently tapering fragment of giant penguin beak sits, unidentified, in the sun.

5 A MOLECULE HERE AND A MOLECULE THERE

A modern compiler, speaking of ambergris, says, "It smells like dried cow-dung." ⁕ G. W. SEPTIMUS PIESSE, *The Art of Perfumery and the Methods of Obtaining the Odours of Plants* (1862)

Its essence, with or without the addition of musk, is mixed with powders, pastes, skin-softeners, and other of those toilet mysteries which men folk are not permitted to inquire about too minutely. ⁕ From an article about ambergris in *Ballou's Monthly Magazine* (1871)

French ambergris trader Bernard Perrin is driving through Normandy when I call him on his cell phone. In the background, combining sometimes to make him completely inaudible, are the sounds of French traffic, the car radio, and Perrin's wife, who talks incessantly.

The Perrins are on vacation. The quiet green fields of northern France flash past in their car window, punctuated occasionally by a village and a volley of cheerful-sounding car horns. But if a call suddenly comes in—informing Perrin that a boulder of ambergris has been found on the Maldivian coastline or somewhere in Southeast Asia—their vacation will abruptly terminate. If the size and quality of the ambergris demands it, Perrin will fly across the world to retrieve it, first buying it from its finder, and then bringing it back to his headquarters on the Côte d'Azur, in the South of France.

On matters such as these, Perrin is politely evasive, and with good reason. "In the past," he tells me, in a thick French accent, "I told somebody that I found ambergris in Maldives and the following day the guy was in Maldives." Within the past year, he finally tells me, he has traveled to the Philippines and the Maldives to collect ambergris. "I travel generally for a quantity of fifteen to twenty kilos," he says matter-of-factly.

⁕   ⁕   ⁕

Almost a year earlier, I had contacted Chandler Burr with a few perfume-related questions. Burr is the full-time perfume critic for the *New York Times* and the author of *The Emperor of Scent: A Story of Perfume, Obsession, and the Last Mystery of the Senses* (2003) and *The Perfect Scent: A Year Inside the Perfume Industry in Paris and New York* (2008).

For his most recent book, Burr had spent a year embedded in the perfume industry, watching as master perfumers built brand-new fragrances for large dynastic perfume houses like Hermès and Coty.

I knew that the highest-quality white ambergris sold for astonishing prices—up to $8,000 per pound in Europe—but I was not sure who was buying it at those prices and what they were using it for. The perfume world is a secretive and impenetrable place, but I hoped Burr would be able to tell me if perfumers were still using ambergris as a component in modern perfumes.

"The scent makers (Firmenich, IFF, et cetera) used to use it," he replied, "and they used to have experts who could assess it, and indeed these were people who had a big responsibility, but no one uses it today (1) because the supply is completely aleatory, up to chance, and you can't build a perfume on a material you may or may not have around and (2) it's an 'animal product,' although this is completely stupid because the whale has already puked this stuff up and gone about its business."

I reminded Burr that ambergris is worth more now than it ever has been, and that perfume manufacturers are rumored to be its biggest users. His response was abrupt but definitive: "No," he wrote, "they only use synthetic ambergris today. Only."

It wasn't the answer I was looking for, but I was willing to accept it. For a year, as I tirelessly searched the local beaches for ambergris, I struggled to answer the most fundamental question: When a pungent boulder of ambergris washes ashore on the beaches of Sri Lanka, Australia, or Indonesia, who buys it and for what purpose do they use it?

And then I called Bernard Perrin, who told me nonchalantly, as green Normandy swirled and flashed past his window, that Chandler Burr was wrong.

*   *   *

At L'École de Pharmacie in Paris in 1820, in a cramped but industrious laboratory, Pierre-Joseph Pelletier and Joseph Bienaime Caventou were about to embark on a new project. The bright Parisian sunlight was angling in through the windows and bouncing off a bristling forest of glass distillation columns and a row of bubbling water baths. On the cluttered workbench, in front of Pelletier and Caventou, sat a small and fragrant piece of ambergris. In the previous few years, the two scientists had made several important discoveries. They were quietly becoming world-famous: there is a monument in Paris honoring their achievements; and in 1970 a postage stamp was issued in France with their profiles on it. Eventually, Caventou would even have a crater on the moon named after him. Together, they had developed new techniques that allowed them to isolate and extract the active compounds from medicinal plants. It was a development that would revolutionize medicine. First, in 1817, they had isolated chlorophyll. A year later, strychnine. A year after that, brucine. Then, in 1820, they had isolated an anti-malarial compound from the bark of the Peruvian cinchona tree and named it quinine—the name came from the Incan word *quinaquina*, which literally means "bark of barks."

Before the discovery of quinine, long strips of dried cinchona bark were ground into a fine powder and mixed with wine into a thick and unpalatable slurry. Malarial patients had to drink large and frequent doses to subdue their fevers, complaining that the mixture got stuck in their throats and made them sick. It worked, but anyone who drank it spent most of the day drunk. Then came quinine. It was a life-saving revolution. Within six years, Pelletier owned a factory that was producing more than three tons of quinine sulfate each year. It was being shipped anywhere in the world there was an outbreak. But in 1820—the same year they isolated quinine—Pelletier and Caventou had turned their attention to ambergris.

Nineteenth-century chemistry was a much more primitive discipline. A description of ambergris in *A Dictionary of Chemistry and the Allied Branches of the Other Sciences* from 1863 reads:

> If good, it adheres like wax to the edge of a knife with which it is scraped, retains the impression of the teeth or nails, and emits a fat odoriferous liquid on being penetrated with a hot needle. It is generally brittle; but on rubbing it with the nail, it becomes smooth, like hard soap. Its colour is either white, black, ash-coloured, yellow, or blackish; or it is variegated, namely, grey with black specks, or grey with yellow specks. Its smell is peculiar, and not easy to be counterfeited.

Perhaps Pelletier picked up the ambergris from the workbench and tested it with his teeth, biting into it like an apple and then holding it up to the pearly Parisian light to look for a bite mark. He carefully weighed and measured it. Together, he and Caventou made notes on its color and odor, calculated its density, and measured its boiling point. After dissolving it in hot alcohol, filtering the solution, and then leaving it to stand, they noticed white crystals forming in bulky and irregular clumps. This was what they were interested in most of all. It was the active compound in ambergris. They named it ambrein. This was not quite the same as isolating quinine. No one was saved by the discovery of ambrein. Almost as quickly as it was completed, their work on ambergris was eclipsed forever by the possibility of using quinine to eradicate malaria. But the known world had become a little larger, and now ambrein was part of it too.

*   *   *

Somewhere in Grasse—"Next to Cannes, next to Nice, in the south of France," explains Perrin—there is a small and darkened storage room. Occasionally, that room is filled with pungent, cotton-wrapped pieces of ambergris worth several million dollars. There are large, rounded pumpkin-size lumps, pale flattened slabs, and long, thin pieces with tapering ends. They have come here from across the world: from the Bahamas, the Philippines, and the Norfolk Islands. For a short time, they are Perrin's. And then, as suddenly as they arrived, they are gone again, sold to the highest bidder.

"You need to buy when you find," Perrin says, explaining the fluctuations in his stock, "because sometimes you don't find some and sometimes you are short. You are short, you have no more, and people are asking, so sometimes you can have stock of one hundred kilos and sometimes nothing. Generally, whenever I find, I sell it very quickly. Sometimes, with some customers, they book in advance and say if you have five good, I will buy it. You can find in India, you can find in Maldives, you can find in Philippines, in New Zealand, in New Caledonia. You can find also in Somalia, but I don't buy in Somalia. You can find in Madagascar."

Almost without exception, says Perrin, the boulders of ambergris in his storeroom were found on the shoreline, washed up by the ocean. In some countries, he employs local agents to represent him, buying ambergris from villagers on his behalf. In other locations, he maintains relationships with people who find ambergris regularly.

"You have some people who are finding," Perrin explains. "They are fisherman, they are used to going on the beach and on the boat, so they are . . ." He pauses, searching for the right word. "I have very good finders, I would say. Regular. Sometimes you don't find at all. You don't buy at all for six months, it depends. Sometimes, you find the big piece. Sometimes you find small pieces. It's not a regular business. We can buy any quantity available. In some countries we have agents, and they will buy for us. Yeah, sometimes we buy one hundred grams, fifty grams, but it will be bought by a local agent."

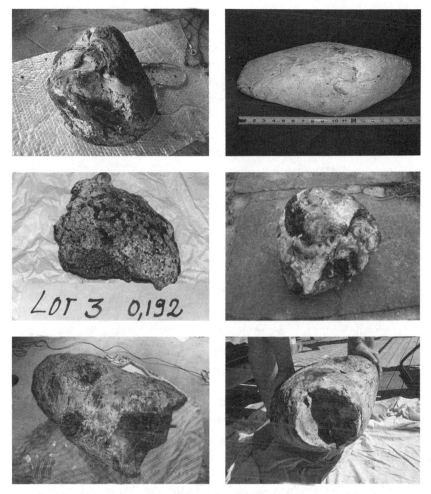

Samples of ambergris from Bernard Perrin's collection. Credit: Bernard Perrin.

A few weeks later, Perrin sent me photographs of several pieces of ambergris he had traded in recent years. They arrived, loaded onto a flash drive, in the mail. There is a photograph of a long and slender piece of ambergris that was found in Bermuda. It looks like it was chiseled from a piece of coastal English chalkstone, and it dwarfs a Marlboro Lights cigarette packet that has been placed carefully on the floor beside it to provide an idea of scale. In another photograph, a piece of ambergris is laid alongside a yellow measuring tape: it is eighteen inches long. It is, in short, not the sort of thing one would miss on a remote beach while searching for ambergris.

Another photograph shows an implausibly large round boulder of ambergris that washed ashore in New Caledonia in May 2007. It weighed, Perrin tells me, 100 kilograms—or around 220 pounds. It is simply huge: an extraordinary, superlative, egg-shaped boulder of ambergris. It is the color of granite, mottled in places with irregular brown and white patches. And it is worth around $1 million. It sits solidly on the floral print of a bedsheet, which has been spread out on a concrete floor some- where in New Caledonia. In the background, two people stand behind it, their legs small and out of focus behind the enormous ambergris that swallows up the foreground.

* * *

In early September 2005, a large piece of ambergris washed ashore in North Carolina. A local woman found it there and contacted Will Lapaz, the proprietor of Eden Botanicals, a northern California–based com- pany that sells wholesale essential oils and aromatherapy products to perfumers.

The same woman had contacted him a few years earlier, says Lapaz, believing she had found a piece of ambergris. On closer inspection, it proved to be a lump of copal—a non-fossilized and immature type of hardened tree resin. But this time, she was certain. She had found am- bergris. "It was after Hurricane Katrina," recalls Lapaz. "It was after a lot of hurricanes. It was that year, late in that hurricane season, so she was looking for it. She's been looking for it all her life, as a kid, for thirty years. And she found it. I said, 'This sounds good, send me a sample.'"

When he received a small, gray, aromatic sample of it in the mail, Lapaz knew it was genuine ambergris. "I sent it off to a couple of people," he says, "you know, old-timers, a French old-timer in the perfumery busi- ness, and he wasn't exactly sure. He had worked with ambergris tincture, but he'd never worked with raw ambergris, so he wasn't sure. I went ahead and I sent it off to other people, who said, 'That's it. Don't break it. Don't do anything. We want to buy it.'"

Eventually, Lapaz bought the ambergris from the woman who found it, whose name he is unwilling to share. After the purchase, he trans- ported it—by means he is equally reluctant to disclose—across the coun- try, from North Carolina to Hyampom, California. "It was several kilos," says Lapaz. Since then, he has been selling it to independent perfumers across the United States, who then use it to make their own ambergris tincture, grinding it up with a pestle and mortar and dissolving it in alcohol. "I sold it off piece by piece," he explains. "I have twenty-five grams left. That's it. Everything went. It took about four years to sell everything I had, and it was all word of mouth. I never advertised it. I never offered it for sale on my website. I never went public with it."

One of the perfumers who bought ambergris from Lapaz was Mandy Aftel. A California-based perfumer, founder of the Natural Perfumers Guild, and the author of *Essence and Alchemy: A Natural History of Perfume* (2001), Aftel has been credited with popularizing a modern and resurgent style of perfumery using only natural ingredients. "I make a tincture of the ambergris, and I put it in my perfumes," she says.

Alongside the French artisanal perfumer Francis Kurkdjian—who charges clients more than $10,000 to create two ounces of custom-designed perfume—Aftel was named one of the world's best custom perfumers by *Forbes Magazine* in 2009. For slightly more than $1,000, Aftel will provide a client with a half-ounce of custom-made, specially designed bespoke fragrance.

In her Berkeley, California, studio, Aftel sits with her clients, carefully building a fragrance layer by layer, in front of a bank of wooden shelves lined with hundreds of glass vials—more than six hundred of them, in fact, filled with different essential oils, tinctures, extracts, pastes, and resins. There are essences of green tea and licorice, and extracts of sarsaparilla and seaweed; there is blond tobacco essence and porcini oil; there is oven essence of Africa Stone, extracted from the petrified and centuries-old excrement of an African mammal called the Cape Hyrax (*Procavia capensis*). And, of course, there is tincture of ambergris, made with the ambergris she purchased from Will Lapaz.

"I don't use it for all of my perfumes because I'm very fickle," says Aftel. "Each perfume for me is a clean white sheet of paper. It's a new narrative, a new experience for me. There's nothing that's in every single perfume of mine. That's not how I work. Although, if something was going to be in every perfume of mine, ambergris would be a good candidate, because I love it."

After spending a significant amount of money on ambergris in the past, and then discovering that what she had bought was not ambergris at all, Aftel was eager to buy genuine ambergris from Lapaz. "It has, I'd say, two functions in a perfume," she explains.

It has a very beautiful, almost hard-to-describe, kind of sweet, ambery, rich, aroma that it brings to a perfume that really makes everything more beautiful. But it also has an effect on the other oils in a perfume. It's got this kind of transformative quality about it. Some of the really magical essences in perfume have a magical quality not just in their aroma profile but in the way that they affect the other essences that you have in the perfume. Ambergris is an absolute star in that department. It creates almost, if you can stretch your mind to imagine this, a kind of shimmery, sparkly effect on the other essences. It moves around inside them and changes the way that they smell, and it makes them just more beautiful. It's the real argument for the naturals, which is: it's more than the sum of its parts.

*   *   *

In the months before I finally smelled genuine ambergris, I devised a plan that I thought would allow me to smell at least a facsimile of its complex and indescribable odor. I would collect a sample of every substance that had ever been used to describe ambergris, and then mix them all together to create a sort of compost. The resulting bouquet, I hoped, might begin to approximate the smell of ambergris.

The idea had a simple practical appeal. And there were plenty of substances to collect: old cow dung, damp leaf litter and freshly turned earth, new-mown hay, seaweed, Brazil nuts, fine tobacco, vanilla, and violets. I began to formulate lists in my head. I would make a potion. On a bright windy morning, I drove to Aramoana and collected wet fronds of sea lettuce from the rocks at low tide. For the first time in a year, as I drove along the harbor road, I looked away from the water, ignoring the ambergris that might be floating there, and stared instead toward the steep green hills. I was looking for a fern copse. I had already located a cliff-top field that was home to a noisy herd of cows. At a later date, I planned to return there, select a cowpat that was neither too fresh nor too old and weathered, pry it from the grass, and take it home. I would add it to the soil, the damp moss, and the seaweed I had already collected. In town, I walked along grocery store aisles, looking for jars of Brazil nuts and affordable cigars to crumble into the mixture. And at night, I wondered if I would have to grow my own violets, carefully watering and nurturing them in pots, only to pluck their petals from their stems and drop them into a growing and slowly liquefying compost.

One day I told my wife I needed to acquire some wood from an old church. She frowned. It was the first time she had displayed anything but unreserved enthusiasm for my growing obsession. After some discussion, we arrived at a compromise: I would combine all of the other ingredients in a container I could carry with me, and then, when I was ready,

I would find an old church, sit in an isolated pew, lift the lid off the container, and bury my nose inside it.

If I had not abandoned my plans—and I'm still mostly relieved that I did—I would have been hoping that my strange compost was also more than the sum of its parts. In fact, I was hoping that, together, its components would smell almost exactly like ambergris.

*   *   *

"Ambergris, as you can imagine, is what we call an animal note," says master perfumer Tony Morris, whose thirty-year career with the Swiss-based multinational perfume and flavor giant Firmenich ended with his retirement in 2001.

His accent is proper and clipped English, made strangely musical by his decades spent in Switzerland. He continues:

> Animal notes—which include musk and a few others, civet and castoreum, which comes from the beaver—these animal notes, they've been used in perfumes because the animal note gives warmth and complexity to a perfume. It's rather like if you have some wines, some of them have animal notes, and it gives this warmth and complexity; and that's the key element of why ambergris and civet and these other materials were used.
>
> The form of the ambergris is rather complex, a lump, you know how it is, colors and so on? It has different colors within the piece. How do you know what is in there until you open it? And you can't open it before you've bought it. Do you know how they test for the quality of it? The key part is the gray part, do you understand? You have to assess how much of the piece is gray, and you do it without cutting it open. Do you know how they would do it? You actually take a little sample, like going down to drill into the Earth to make an oil well, and they take a sample of it going through. Of course, the shape of it can be confusing because the white part or the gray part can vary according to the shape. In some parts you make a little insertion with a special tool and you take a sample out, a little fine sample, and you find, "Oh, there's gray of this amount," so you make an offer to how much to buy. That's how they do it, or used to do it.

*   *   *

In 1792 a rowdy enthusiasm was sweeping across revolutionary Paris for a new invention: the guillotine. Used for the first time on April 25, 1792, to dispatch Nicolas Jacques Pelletier, a highwayman, it was considered a modern and humane method of execution. Two years later, thousands had shared Pelletier's fate. Heads had rolled. Louis XVI was executed by guillotine in January 1793. Marie Antoinette, found guilty of treason, followed him nine months later.

So bloody was this business that in Paris, in front of the Saint-Antoine gate, an aqueduct was constructed to collect and redirect the blood of the guillotined. And at Metz to the east, the heads of the guillotined were placed on the tops of their houses.

"At elegant dinners," wrote the French historian Imbert de Saint-Armand a hundred years later in *Marie Antoinette and the Downfall of Royalty* (1891), "a little guillotine is brought in with the dessert and takes the place of a sweet dish. A pretty woman places a doll representing some political adversary under the knife; it is decapitated in the neatest possible style, and out of it runs something red that smells good, a liqueur perfumed with ambergris, into which every lady hastens to dip her lace handkerchief. French gaiety would make a vaudeville out of the day of judgment."

In 2005 Elisabeth de Feydeau, a Sorbonne-educated historian and a professor at the Institut Supérieur International du Parfum de la Cosmétique et de Aromatique Alimentaire, joined forces with artisanal perfumer Francis Kurkdjian. Their goal was to re-create a 200-year-old fragrance that had been made by perfumer Jean-Louis Fargeon for Marie Antoinette, the queen of France. The result was a fragrance called Sillage de la Reine, which translates approximately to "Wake of the Queen." A thousand bottles were made, containing 25 milliliters of perfume and priced at almost $500 each. For high rollers, ten Baccarat flasks, each filled with 250 milliliters of the honey-colored fragrance, were sold for $11,000 apiece.

I e-mailed de Feydeau, who rediscovered the formula for the perfume in a notebook belonging to Fargeon while researching her book *A Scented Palace: The Secret History of Marie Antoinette's Perfumer* (2006). When I asked if it contained ambergris, she responded, "Of course, we used ambergris."

In fact, Sillage de la Reine contains all natural ingredients: a spectrum of floral notes—of rose, iris, jasmine, orange blossom, and tuberose oils—tethered to the base notes provided by Tonkin musk and ambergris, in proportions that de Feydeau refuses to share.

"I thought, 'Well, we can play music from the eighteenth century, why not re-create the perfume from the eighteenth century?'" explains de Feydeau when I speak with her. "I had the book of perfumery of Jean-

Louis Fargeon, the perfumer of Marie Antoinette, and by the other hand,
I had the orders of the queen, conserved in the National Archives. I knew
the tastes of the queen on matters of perfumery. There was a recipe called
les mille fleurs, the thousand flowers. When you read the recipe, you have
aromatic notes; you have, of course, flowery notes; and, to give this trail,
to give to the perfume a trail and to give the power, you have animalic
notes: the ambergris and musk."

* * *

Back in Grasse, another huge gray lump of ambergris has arrived from
the other side of the world. After Perrin wraps it carefully in cotton, he
will lovingly stow it away in his storeroom with the others. They wait
in the darkness, like broken statuary. Another piece has been sold to a
wealthy buyer. It is removed from storage and prepared for the final leg
of its journey. It might travel to the Middle East, where it is still used as
an aphrodisiac and a tonic.

In *The Iron Wall: Israel and the Arab World* (2001), author Avi Shlaim recounted a moment during the strenuous September 1978 Camp David
peace talks between President Carter, Egyptian President Anwar al-Sadat,
and Israeli Prime Minister Menachem Begin when Sadat's astrologer, a
religious mystic called Hassan Tuhami, distributed pieces of ambergris
to the Egyptian delegates, "telling them to dissolve it in their tea, for it
would give them the stamina to confront the Israelis."

Reportedly, although several delegates followed Tuhami's advice,
"Boutros Boutros-Ghali declined the offer."

If not somewhere in the Middle East, Perrin's ambergris might end up
in Asia. "Sometimes Singapore," he says, "but I don't know the market."
Alternatively, and most often, a typical piece of ambergris will travel just
a few miles away, destined for one of the numerous world-renowned perfumeries clustered around Grasse, on the lower slopes of the French Alps,
overlooking the Mediterranean ocean.

"It is mainly used by well-known brands like Chanel, Guerlain," says
Perrin, "but they will never buy from a finder. They will buy from a spe-
cialist."

For more than two hundred years, Grasse has been considered the
perfume capital of the world: Galimard, established 1747; Guerlain, 1828;
Molinard, 1849; Fragonard, 1926. A small city approximately twenty
miles from the coast, with a population of around fifty thousand residents, Grasse is surrounded by rose, lavender and jasmine fields. Each

year in July and August, their petals are still harvested, collected like delicate fruit, and transported in sacks to nearby perfume houses like Chanel for processing. Other lesser-known companies in Grasse specialize in producing the fragrant materials for perfume manufacture. "Oh, you have Mane," says Perrin, "you have Robertet, you have Charabot, you have many, many companies."

Respectively: Mane, established 1871; Robertet, 1850; Charabot, 1799.

For most of that time, while tons of lavender, jasmine, and rose petals were being harvested from the surrounding fields, pieces of ambergris were arriving on the Côte d'Azur too, from across the world. There, the ambergris was made into a tincture. "It's very complicated to actually make it into the infusion, as they call it, and then the infusion is used with the alcohol before you put the perfume essence in," master perfumer Tony Morris told me from his home in Geneva.

I ask Perrin if he has ever sold ambergris to established perfumeries like Chanel (est. 1909).

"Indirectly, yes," he says, "indirectly. These people, they have their own in-between men. It's a bit complicated, but we have sold; in France we are selling through an agent. You know, to sell to Chanel, Guerlain, you need a special contact. So we have to sell to special perfumers who are in contact with Chanel. They will buy ten, or twenty kilos at a time, but they are very selective. They will buy top quality, and they will choose from the stock you have. So, they will buy maybe fifty kilos, plus. Regarding the prices, yes, they will buy, it depends, between ten and fourteen thousand Euros a kilo. Only top quality. Top quality. They don't buy only from myself. They buy from other companies. This is a delicate subject because you're asking me my turnover. I can't give you information about the price because some people, I know that you have contacted Adrienne from New Zealand, she will sell per gram, she will sell $20 per gram, but she's a small business. One gram by one gram. We sell by the kilo."

It was a revelation to hear Perrin nonchalantly explain in his thick French accent how he obtains gigantic boulders of ambergris from around the world, and then sells them to some of the most established and successful perfume houses in history.

*   *   *

Ambergris is valued for two unrelated properties, both of which explain its widespread use by perfumers. First, it is a powerful fixative, stabilizing fragrances so that they last much longer on the skin. Elizabeth I perfumed her gloves with ambergris not so much because she enjoyed its fragrance but because it lasted for years, surviving wash after wash. A piece of ambergris will retain its odor for three hundred years. This particular property was a source of curiosity for centuries.

"Mr. *Boyle* had a Pair of Gloves in which the Smell of the Ambergrease continued for above twenty Years, and yet smelt very strong whenever he opened them," the Dutch physician Herman Boerhaave told University of Leydig students in a 1745 lecture.

> I formed a Pastil of Musk, Ambergrease and Civet mixed in a certain Proportion, and placing it in a wooden Chest not accurately closed, it has continued there for above thirty Years, still retaining its Fragrancy in a great measure: yet it does not smell at all times equally strong; for when the Air is moist or suddenly changed, it smells more intensely, whereas at other times it is very weak. The whole Affair seems to follow from the Minuteness of the odoriferous Particles, which no One could ever discover even by the Microscope, either in the Flower itself, or flying off from it in the Air; no one could ever perceive them either by the Taste or Touch: but at the same time in these wonderful minute Particles, there is Force enough to briskly affect the Nose.

The second, and equally important, property of ambergris is its indefinable odor, which people have always struggled to describe in a meaningful way.

Despite failing the basic challenge of description, perfumers use a specific adjective to describe the nonspecific odor profile of ambergris: animalic—referring to the coarser and less refined notes of a fragrance that seem to lurk darkly beneath the lighter and more appealing floral tones. Since Pelletier and Caventou first isolated ambrein in 1820, chemists have deconstructed it even further, dismantling it atom by atom to find the source of its complex odor. Ironically, ambrein is odorless. But it holds within its structure all the necessary building blocks to make other aromatic compounds, each with unique and individual odor profiles.

"The ambergris molecule," says Dr. Charles Sell, "it's a terpenoid."

It is midmorning in Kent, sixty miles to the southeast of London, and Sell is in his office at one of the United Kingdom locations of Givaudan, one of the world's largest fragrance and flavor manufacturers. "The terpenoid chain is cyclized at both ends," explains Sell, the author of a book titled *A Fragrant Introduction to Terpenoid Chemistry* (2003). "You've got a decalin-type ring at one end and you've got a cyclohexane ring at the other end, and the chemistry that goes on breaks the chain that links the two. So, you break it down into the two smaller fragments, which are then volatile enough to reach the nose and smell. But there are lots of different molecules that are present in the brew that you get."

For instance, take ambrein—the odorless active compound in ambergris, isolated by Pelletier and Caventou in 1820—and trim it a little, reducing the long tricyclic chain of carbon atoms to almost half its size, and the result is a different compound altogether. It is now gammadihydroionone, a shortened monocyclic compound. It still closely resembles one-half of ambrein. But it is no longer odorless. Instead, this abbreviated portion of ambrein smells strongly of tobacco. If a methylene group—one carbon and two hydrogen atoms—are added to ambrein, occupying the position next to the single oxygen atom that dangles at the end of the carbon chain, the result is yet another compound: 2-methylene-4-(2,2-dimethyl-6-methylenecyclohexyl) butanal.

And this molecule—altered, but almost imperceptibly—smells like seawater. From no odor at all, to tobacco, and then to seawater: a transformation. A few subtle changes have altered the shape of the molecule completely. Its three-dimensional structure is different now. It folds over on itself in new ways. It occupies a different shape in space. When gamma-dihydroionone binds to olfactory receptors in the nose, it sets off a relay of nerve impulses in the brain, which then registers the smell of tobacco. Add a carbon and an oxygen atom to it, and it binds differently to those receptors. Suddenly, it is seawater instead. If gamma-dihydroionone now undergoes a process called cyclization, adding a second carbon ring where before there was only one, the result is another new molecule: alpha-ambrinol. Shortened and more compact than its predecessor, it has its own unique odor profile, which is described as moldy, animal, and fecal.

Finally, we return to the remnant of the original ambrein molecule, the portion that was trimmed away and gave rise to the gammadihydroionone molecule, and the succession of other degradation products that came after it. The addition of a third carbon ring to this shortened chain, a pentagonal ring to join the two hexagonal rings already present, produces a tricyclic compound called 3a,6,6,9a-tetramethyldodecahydronaphtho[2,1-b]furan, or more simply, naphthofuran. And this molecule, which looks in so many ways like the ones that preceded it, smells like ambergris.

Naphthofuran—also known as ambergris oxide. This nondescript molecule is the final outcome of a decade or more spent at sea. As ambergris slowly transforms from a viscous and unpleasant-smelling waste product to a pungent and resinous light gray boulder, ambrein has degraded to become ambergris oxide and a rich fragrant mixture of other organic molecules. This is steady entropy and inevitable breakdown: after years of oxidation by equatorial sunlight, almost completely submerged in seawater, the atoms are stripped away, added at one location, and then removed from another. Double bonds become single bonds; chains become rings, structures change, the shapes of molecules are altered, and new compounds emerge.

"The compound that's the most distinctive one for ambergris, and probably the one that has the most important role in it, is the naphthofuran," says Sell.

> It's got lots and lots of trade names. The Givaudan trade name for it is Ambrofix. The Firmenich trade name is Ambrox. Henkel calls it Ambroxan, and there are lots of other trade names, but that molecule is the one that, if you smell the individual pure molecules, this is the one that is most characteristic of ambergris and the one that's impossible, or nearly impossible, to mimic from outside that odor area. There's nothing else. We've got synthetic molecules now that mimic it, but in terms of the odor character, that's the unique bit of ambergris. If you look at the descriptions that people give, one of the breakdown products is described as briny ozone, another one will be described as tobacco-like, but the naphthofuran, the only label we can put on it is ambergris, because there's nothing else quite like it.

If chemists try to build something like ambergris oxide in the laboratory, they fail on a fundamental level. Sell explained later via e-mail:

> Another fascinating aspect of this is the stereochemistry, that is the arrangement of atoms in space. If you were to just clip the five-membered ring onto ambrinol (how often I have wished I could do things like that to molecules) the oxygen atom would be on the wrong side of the two six-membered rings. That is, in the way we normally draw the naphthofuran, the oxygen would be above the plane of the paper instead of below it. This molecule is known and is very much weaker than the natural version with the oxygen down.

People have struggled for centuries to describe the odor of ambergris. Their failure to do so, says Sell, has a chemical explanation:

> Practically everything that you smell in nature is a complex chemical mixture. Ambergris doesn't have any more components than jasmine oil or rose oil. They're all in the hundreds of components. I think the reason why people will have difficulty describing the smell of ambergris is that there are no reference points for smell. For color we have red, orange, yellow, green, blue, indigo, violet—the colors of the rainbow—and they're all associated with specific wavelengths of light. So you've got a simple primary color refer-

ence when you're trying to describe a color. With smells, every smell is different, and every smell—whether it's a single molecule or a mixture of a thousand different molecules—it will create a pattern on the olfactory bulb in the brain. Smell is then recognition and interpretation of that pattern on the olfactory bulb.

In other words, people struggle to describe the smell of ambergris because there is simply nothing else that smells quite like it. Only ambergris smells like ambergris. It is singular. To complicate things further, the ambergris smell is complemented, modulated, and adulterated by the presence of the other breakdown products: gamma-dihydroionone (tobacco), 2-methylene-4-(2,2-dimethyl-6-methylenecyclohexyl)butanal (seawater), and alpha-ambrinol (mold, animals, and feces). But the naphthofuran is a hurdle that even the most imaginative of us cannot leap. There are no reference points. It is like a single, remote point on a map with no landmarks anywhere by which to find it. Describing its odor to someone who has never smelled it is like trying to describe a rainbow to a blind person. Words are inadequate, but they are all we have.

*   *   *

Chemists have gone to great lengths to try to produce something in the laboratory that smells like ambergris. Despite the constraints of stereochemistry, they have succeeded to some extent. The results—patented molecules with trade names like Ambermore, Cetalox, and Synambrane—are used by perfumers to provide the animalic tones of ambergris.

"In my drawer here, I've got a couple of dozen molecules that are all in that odor area," says Charles Sell. "There's quite a range of different compounds. They're mostly fairly big compounds. Obviously, to get the same odor, you have to act on the same receptors, and so the naphthofuran is largish; and in terms of fragrance molecules, it's one of the bigger ones, so the other things that smell like it tend to be the bigger molecules and mostly are ethers or alcohols, very often cyclic ethers."

The process is simple enough: the vital starting material is clary sage, a common herb with long stems and flowers that grow in thick brush-like sprays. It was a technique first developed by the Swiss chemist Max Stoll. In 1953 Stoll traveled from the Firmenich headquarters, in Geneva, to Los Angeles, to accept the prestigious Fritzsche Award from the American Chemical Society. He had earned the award for replicating the structure of ambergris in the Firmenich laboratories four years earlier. The compound had since been patented and christened Ambrox. It revolutionized the perfume industry. And it is still used frequently in perfumery today.

Stoll's achievement was so notable that he spent the two weeks before he accepted the award crossing the United States, visiting universities and chemical research centers as he went.

"To learn its chemical constituents," reported the *Washington Post* in March 1953, "Dr. Stoll started with a block of natural ambergris. He conducted a long series of chemical processes in an attempt to duplicate its chemical parts and structures to produce man-made ambergris."

Months later Stoll and his coworkers had created a compound that was identical, they thought, to the chemical structure of natural ambergris. When they subjected it to a battery of tests, it passed all of them but one: the all-important smell test. "Neither he nor his associates could detect the odor of their product," reported the *Post*. "It was put on a shelf and further efforts were made to reconstitute ambergris."

In fact, their noses had failed them. Sitting on the shelf in the laboratory was a vial of what was later named Ambrox—pure, synthetic naphthofuran. It had evaded chemists for decades, and Stoll had succeeded. But no one had noticed. For six months, it waited on the shelf above him, as he sat at his workstation below, correcting his formulations, and checking and rechecking his impenetrable calculations.

"One day the firm's technical director, Dr. Roger Firmenich, came into the laboratory," the *Post* article continued. "'You've got it!' he exclaimed. 'The whole room smells amber.' Dr. Stoll and his associates just stared. They hadn't realized till then, he said, that their noses were 'intoxicated' by the odor they had been seeking."

* * *

In the past few decades, perfume houses have lowered and then lowered again the formulation costs for their fragrances, and they do this by using more and more synthetic compounds. Only a handful of houses remain that are willing to pay for the most difficult to obtain raw materials, like ambergris. Most other houses prefer to use synthetic ambergris compounds like Ambrox, which, as Luca Turin wrote in his comprehensive book with coauthor Tania Sanchez, *Perfumes: The Guide* (2008), "smell nothing like the natural material."

In fact, synthetic compounds only simulate ambergris, sometimes successfully and sometimes much less so. Ambermore is not ambergris; neither are Ambrox nor Synambrane. They impersonate certain olfactory and fixative aspects of ambergris. But they do not replicate it. Comparing genuine ambergris with a synthetic version of it is like comparing an original painting by Van Gogh with an inexpensive print of the same canvas.

In a 2005 article for the *New Yorker* magazine, perfume critic Chandler Burr profiled a French perfumer whose nose was so sensitive that he could smell a vial of jasmine essence and identify not only the country in which the flowers were grown but whether the machines they were processed in were made of aluminum or stainless steel. The 2009 *Forbes Magazine* article that named Francis Kurkdjian one of the world's top custom fragrance designers recounted his response to a client who requested a perfume that smelled of vetiver grass: did the client want the fragrance to smell of Haitian, Chinese, or Javanese vetiver?

This presents an interesting paradox: some of the perfumers who use synthetic ambergris have noses that are so discerning—able to detect the infinitesimal differences between Moroccan or Italian jasmine—that they must be fully aware of its inferiority. And yet, in order to reduce manufacturing costs, they use it anyway.

If replicating a molecule in the laboratory that takes decades to form in the natural world sounds too good to be true, that's because it is. First, although ambrein is the active ingredient of ambergris, it only makes up between a quarter and a half of ambergris that washes ashore on remote beaches across the world. Another third or so is made of a compound called coprosterol; and there are a range of other organic compounds present, including ketones, free acids and esterified acids, and pristane. Perhaps these other compounds also contribute to the complex odor that characterizes ambergris. Second, ambrein degrades slowly. Before finally forming the naphthofuran, it first forms a compound that smells like tobacco, which degrades to form another compound with a strong seawater odor, and then another, which smells like mold, animals, and feces.

A substance that requires decades to slowly degrade, changing from one thing into another—from odorless ambrein to indefinably fragrant naphthofuran—is never fully one thing or the other. Instead, it is a combination of both starting point and end product, along with a dynamic and constantly evolving range of all the fragrant by-products created along the way. This is what gives ambergris its power and its complexity. It is indefinable because it is many things. And it is always changing and becoming something new. The journey takes decades and is unique for each piece of ambergris—none are quite the same. Ambergris cannot be re-created simply by processing several tons of clary sage in an industrial silo. But something that is similar enough—a simulation, a specter, a suggestion of the real thing—can be made.

"The synthetic versions have nothing to do with the real ambergris," Bernard Perrin says indignantly from his car, above a sudden chorus of Norman car horns. "Of course, the Americans: ecology, Green Party, blah, blah, blah. They don't want to use any animals products [yet] they eat a lot of meat. Some of them are against having animal in the perfume. The traditional perfumery, the French, old French, they will use ambergris. They will never change. You cannot change a perfume like Chanel Five. The success is nearly one hundred years. It was launched in 1925. They will never change the formulation."

In some quarters, Perrin says, it is a matter of French pride to use ambergris.

> Chanel Five or Guerlain perfume like Shalimar, they will have ambergris. French perfumery is more traditional. They will buy for the quality, and they're used to use such product. They will never change. American perfumery, they are only thinking about making profits, so they will use synthetic ambergris. Ambroxan. They will use Ambroxan because they are thinking about the price. It's a different style.
>
> It's like in food, eh? If you compare a good French restaurant, the cook will know what to buy, how to cook it; and if you go to a lousy restaurant, maybe he will serve you a steak, but he will serve you in a lousy preparation. You can substitute butter by margarine, but it's not the same. Ambergris, it's like your wine, you have different wine, you cannot compare all Bordeaux to cheap wines from . . . I don't know.

*   *   *

Not long after speaking with Bernard Perrin, on a cold and flinty day on Long Beach, I trudge along the margins thinking again about the structure of naphthofuran: a cluster of three aromatic rings. Compared to the other molecules, with their long carbon chains and their dangling oxygen atoms, it looks somehow more complete, more reasonable. This is what the brew of other organic compounds are destined to become after silent years of degradation. Time snips away the excess parts atom by atom, refining and trimming the structure, slowly turning carbon chains into rings.

It is one of the only uncomplicated things about ambergris. When I e-mail Charles Sell to ask him precisely how ambergris stabilizes fragrances, he replies: "The 'fixation' effect is probably due to affinity between the ambergris molecules and the molecules of more volatile ingredients, thus producing a deviation from Raoult's law. In other words, non-bonded interactions between the two sets of molecules result in the more volatile molecules sticking to the ambergris and thus to the surface longer than they would if no fixative was present."

With noisy gulls clamoring above me in the wet air, I crouch close to the sand and try to look past every little rounded pebble and empty crab shell to find the single piece of ambergris that I'm sure must be here somewhere, hidden in plain view among the tidal drifts of shingle. I move slowly across the multicolored scree, pebbles crunching beneath my feet. The week before, I had walked along Tomahawk Beach in the wind, feet sinking into the fine white sand, smelling the briny wind that blew on my face. I had found nothing there either. Making my way along the beach, I had picked up any stone that looked unusual or had strange striations on it, or a surface marked with black or yellow flecks. I had rubbed a fingertip over its wet and pockmarked surface, brought it to my nose and inhaled, and then stored it carefully in my pockets, like a fruit picker.

As the tide recedes, I think again of the files Perrin had sent me—photographs of massive blocks of ambergris collected from across the world and transported to his storeroom on the sun-drenched Côte d'Azur. There, on the other side of the world, wrapped in their cotton shrouds, they wait for a buyer, like ghosts in the darkness.

6 CLOSE ENCOUNTERS OF THE AMBERGRIS KIND

It is possible! Only some gramme. I can send to you by simple mail, to inlay an ambergris in a souvenir. But if you want to purchase everything, then will be to arrive you on Ukraine and take away. To me from the Arabic Emirates arrived and took away. * Personal e-mail from VALENTYN, an ambergris trader based in Odessa, Ukraine (April 2010)

I am finally, after these long months, holding a small rounded piece of ambergris in my gloved hand. Its black surface glitters under fluorescent light. Sitting in the hollow formed by my cupped hand, it looks like a bituminous little pebble, of the sort scattered over a resurfaced road by a retreating road crew.

Outside, it is early autumn. The sky is gray and leaden. Half an hour earlier, I had walked across a large well-kept apron of lawn in front of the Otago Museum, beneath tall trees that are beginning to lose their leaves. At the museum's busy reception desk, I am met by Cody Fraser, the friendly natural sciences registrar, and together we walk down a short corridor toward an unmarked swipe-entry door. Fraser swipes us in. I am now, she says quietly, "behind the scenes." On the other side of the door—in cabinets, drawers, and shelves, and arranged haphazardly on tables and elsewhere—are all the items in the museum collection currently not on display: stored in the dark and protected from fluctuations in temperature and humidity. At any time, Fraser tells me, only 1 or 2 percent of the items in the collection are being displayed. The rest are waiting here, in darkness, packed snugly in camphor to ward off the destructive attention of insects.

When Fraser switches on the lights, I discover that I am standing next to a large blue plastic model fish, surrounded by fossils and scattered pieces of rock. A stuffed possum is lying stiff-legged on its side on a nearby work surface. A stag's head, complete with antlers, stares down from the wall, like an indignant coat rack. Fraser apologizes for the smell of camphor. "Some people start to gag when they come in here," she says matter-of-factly.

We walk past banks of wheel-mounted beige storage units that stretch from floor to ceiling, their wheels running along metal tracks in the floor. As Fraser carefully pushes shelves aside in search of ambergris, I catch quick glimpses of butterflies, bones, and large polished pieces of quartz. I see a display of stuffed birds, frozen in flight and perched attentively on branches. Pushing aside another shelf unit, Fraser stops. Her search for ambergris is over. It has been significantly more successful than mine. She points to a shelf at hip height: two small brown cardboard boxes sit on it. Inside each one is a small piece of ambergris. "If you want to hold it, you can," Fraser says, "but you have to wear gloves."

The first piece I examine is the smallest, nothing more than a fragment. It is mostly black, its surface marbled with thin gray seams that resemble a network of veins. The box is lidless. The ambergris sits inside it like a strange dark egg. A little strip of paper like a fortune cookie fortune informs me in a black spidery script, written more than a hundred years ago, that the ambergris was presented to the museum in 1899. On my request, Fraser places it on a digital scale she has retrieved from among the scattered taxidermy. It weighs a little over two grams. It resembles one of the several hundred little black fragments of volcanic rock I have collected from local beaches in previous months. I marvel at the journey it has completed so that I can hold it now, between my finger and thumb, like a charred almond shell. Museums are orderly places, but ambergris is the product of a long chain of random and unpredictable events, the result of a series of improbabilities. It is simply an unlikelihood. Nevertheless, I am holding a piece of it.

Standing next to me, Fraser watches closely. I can feel her reading my notes as I write them. She's watching my hands as I bring the ambergris to my nose and breathe deeply in. It is the moment I have been waiting for. The months of searching have come to this. Eyes closed, I anticipate the complex array of odors I have read about. But it smells of camphor. In fact, everything in the room smells of camphor. The butterflies in the stacks smell of camphor. The stuffed birds frozen mid-flight smell of camphor, as do the model fish, the polished shells, and the stuffed possum. The mounted deer head, high on the wall above us, smells of camphor, and so do its antlers. I am willing to guess that Fraser smells of camphor too.

I ask her how often people contact the museum with objects they think are ambergris. "Not very often," she says with a shrug. "It's happened about three or four times since I've been here, which is five years."

In most cases, Fraser can correctly identify the object for its finder. In every instance, it has been a disappointment, something else that has been mistaken for ambergris. "It's usually a piece of rotten flesh or a stone," she says. The museum has honorary curators in geology and marine sciences who can help if she's unable to identify an item. "Material that comes from whales all tends to smell the same," she explains, "a mixture of fattiness and blood. It's not pleasant but not too horrible either." I smell the ambergris again while Fraser watches, surreptitiously trying to detect either fattiness or blood, or anything else for that matter, beneath the layers of camphor. "New bones," she says cryptically, "smell the same as old bones."

The second piece of ambergris is larger: rounded and golf ball-size. It is black and shiny; its surface has a satisfying patina to it. Gifted to the museum in 1915, its entry in the museum's digital database simply reads: "Register entry: Dec. 31. Ambergris Catlins District," referring to a picturesque, and particularly wild and rocky, stretch of coastline to the south.

At some point in the distant past, a small fragment of this larger piece of ambergris was sheared cleanly away, leaving a flat surface that provides a glimpse into its interior, which also smells, incidentally, of camphor. Where it is broken, the ambergris is lighter in color, like ash. It is sandy in texture, with clearly defined strata running through it like mysterious growth rings. Tiny squid beaks embedded in the ambergris during its formation appear now in cross section as little round bubbles frozen in the strata. This piece weighs eleven and a half grams and is worth several hundred dollars. I smell it halfheartedly but detect only eye-watering amounts of camphor.

I place it back in its box. This visit has not been reassuring at all. I had been hoping that, once I had finally seen ambergris firsthand, I would somehow be armed with vital new information that would allow me to glance at the unsorted till scattered on the beach and sort through it from a distance, visually, without even taking another step. But these two pieces of ambergris were unremarkable. And their prized scent had been lost. Had they ever been typical? Fraser didn't know, and neither did I. Had a century in storage changed and transformed them in unexpected ways? It was impossible to know. One thing was certain: I had walked past millions of objects that look just like these in the last few months. Millions and millions of little black stones that had been carried onto the sand by a high tide and stranded there in the rain among the seashells and the rubbery green folds of kelp.

* * *

By this time, I am willing to take any advice, no matter how obscure and questionable its source. One day, almost hidden in the British history archives of the Institute for Historical Research, I find a set of 400-year-old instructions for finding ambergris, issued by Sir Francis Godolphin, an English Member of Parliament. Among his numerous other stations, Godolphin was the high sheriff of Cornwall. According to historian Richard Carew, writing in *The Survey of Cornwall* in 1602, Godolphin's "zeal in religion, uprightness in justice, providence in government, and plentiful housekeeping, have won him a very great and reverent reputation in this country."

In 1604, at the end of winter, Godolphin had acquired a small piece of ambergris from the rugged Cornwall coastline of his constituency. The land on which it was found was the estate of his neighbor, Viscount Cranborne. For a punctilious English gentlemen, this could be a problem. It was a matter of no small propriety to resolve the issue.

"In my last I signified unto your lordship the answer of Walter Daniell of Truroe," Godolphin wrote to Cranborne on Valentine's Day, 1604, "how he once had a small quantity of ambergris found within your manor of Ellinglase and how he refused to confirm with his voluntary oath that he had therein set down his full knowledge."

What Godolphin wrote next interests me even more:

> I have now attained a piece found to the westward of your land, weighing scarce two ounces. The party from whom I had it, alleging his skill to be small in the manner to find it, says that such as are skilful covet the wind between them and the places they search and soonest discover it by the scent, as it is said the foxes by the smell find it. Such as are not perfect in the knowledge of it make their proof by casting a little on the coals, whereon it will fume as frankincense.

There was more, as Godolphin then proceeded to wax lyrical, observing that ambergris was "richer in value than the finest gold being thrown out of that great glassy meadow of the sea." But the lines that stay with me are those that came before. For months after I first read them, every time I walk the high-tide line, I will think of them again: *Such as are skilful covet the wind between them and the places they search and soonest discover it by the scent, as it is said the foxes by the smell find it.*

I am struck not by the analogy in particular, but more by the thought of Godolphin, the high sheriff of Cornwall and lord of a considerable estate, crouching over the sand on the wild English coast, breathing in the cold tidal air as it bends the tussocky grass, trying to discern whether the smallest piece of ambergris is hidden nearby in a sandy hollow. I had thought myself a strange sight—waddling, raincoat-wrapped, child-wearing—as I walked through the endless rain on Long Beach. Now I know I am not so strange—or, at least, I am still strange, but I am now at least in better company.

*   *   *

This is how it happens. A man has managed somehow to clamber atop the whale. He sits in the morning sun, proudly astride its slippery silver flank. It is early, but the air is already steamy with the tropical heat. Below the man, on the sandy beach, a crowd watches impassively. The surf breaks against the dead whale's long box-like head. The man surveys his kingly domain. There are more than a hundred people assembled on the beach. They stand in the surf and push against the 50-foot-long whale carcass, milling around hopefully on the sand. With every passing minute, the crowd grows larger. The sweet, nauseating smell of decay hangs over it all.

The man atop the whale is not Louis Smith. This is not July 1891. It is July 2010. In fact, the self-appointed leader of operations—who now points and issues instructions to the crowd—is a Sri Lankan villager. He's sitting astride the lean carcass of a dead sperm whale, which has finally come to rest, perpendicular to the shore, on Manpuriya beach in Mundalama, in the North Western Province of Sri Lanka.

The whale had washed ashore a day earlier, near the small fishing village of Puttalam on the island's west coast, sixty miles north of Colombo, the capital city. The tide had brought it in from the ocean and rolled it unceremoniously up the sloping beach. Its arrival on the shoreline filled the modest village with excitement. Inevitably, people soon began to wonder if the carcass held some ambergris.

The crowd presses closer. A bright-yellow mechanical backhoe is maneuvered inexpertly across the sand, cleaving the crowd into two watchful camps. The bucket is raised into the steamy air. The crowd waits. The surf breaks. And the bucket is finally lowered, through the whale carcass, separating the pale stiffened flukes from the rest of it.

Out it comes: little black-brown boulders of ambergris, rounded like eggs, delicately marbled with tea-colored irregular seams. Photographs of the haul accompany a report of the incident in the *Daily Mirror*, an

English-language Sri Lankan newspaper, under the headline "PEOPLE CUT UP WHALE SEEKING AMBERGRIS." In one, a long-fingered brown hand cradles an apple-sized lump of ambergris in front of a ragged saffron-colored dress. In another, a smiling woman stands next to a group of frowning and unsure villagers, holding a broken plastic bucket toward the camera: in the bottom, several large pieces of ambergris. On the beach, the whale is slowly dismantled in front of the watchful crowd. Piece by piece, the ambergris is removed from its intestines, and afterward the carcass is buried beneath the sand on Manpuriya beach.

*   *   *

The whale that washed ashore on Manpuriya beach had looked lean and sickly. Scientists from Sri Lanka's National Aquatic Resources Research and Development Agency told reporters from the *Daily Mirror* that the whale had weighed an estimated ten tons. An adult male sperm whale measuring fifty feet in length should weigh closer to thirty-five or forty tons. This specimen was underweight. From the 1823 edition of the *Encyclopaedia Britannica*: "It is observed, that all those whales in whose bowels ambergris is found, seem not only torpid and sick, but are also constantly leaner than others; so that, if we may judge from the constant union of these two circumstances, it would seem that a larger collection of ambergris in the belly of the whale is a source of disease, and probably sometimes the cause of its death."

During the whaling era, whalers knew as much without having to refer to a copy of the *Encyclopaedia Britannica*. The intestines of underweight sickly looking whales always required a careful and thorough inspection. A single boulder of high-quality ambergris could be worth more than all the barrels of whale oil collected during the long months spent at sea.

*   *   *

The efforts of the Sri Lankan villagers on Manpuriya beach in July 2010—communally disemboweling a whale carcass for its ambergris—are not uncommon. The grisly practice is unlawful in many countries; nevertheless, it happens all the time. "The most likely way that you would find [ambergris]," says Anton van Helden, the marine mammals collection manager at the Museum of New Zealand Te Papa Tongarewa in Wellington, "would be from cutting it out of a sperm whale. We've had a number of examples, particularly up in Northland, the west coast, of animals coming up and being discovered, found by DoC [Department of Conservation], and they've been disemboweled. But that's illegal."

When I first spoke with John Vodanovich of the Dargaville Beach Mafia, he had mentioned the practice too. "Did you know that people chop up whales and get it out?" he had asked me. "That's what happens on our beach. I know of a couple of whales that have been chopped once they've washed in on the beach. They've gotten out fifteen-kilo hunks. One on Muriwai was forty-eight kilo." He had paused, and I could hear him slowly calculating its value under his breath. "Times ten," he said to himself quietly before continuing, "that would be $480,000, eh? The guy got paid $76,000 for it. He was ripped off big-time."

But Vodanovich disagrees with van Helden about the frequency of these events. Most of the ambergris bought and sold commercially, he says, is found on the shoreline, and not cut from whales. Full-time ambergris collectors spend months of each year hiking to some of the wildest and most remote locations in search of beach-cast ambergris. Illegal or otherwise, professional collectors cannot afford to wait for whales to strand and die. It happens too infrequently.

A sperm whale stranding is a rare event, agrees Laura Boren. The national marine mammal coordinator for the New Zealand Department of Conservation, Boren says that between 1873 and 2009, approximately two hundred sperm whale strandings have been recorded on the New Zealand coastline. "This rate of stranding," she explains, "is what we would consider regular, in that we can anticipate that at least one a year will strand, but they do not strand frequently."

In just two of those two hundred sperm whale strandings, people had cut into the stomach of the whale, in an attempt to harvest ambergris. "It is important to note," writes Boren via e-mail, "that all stranding events and incidents of human interference (e.g., cutting open the stomach to retrieve ambergris) are likely to be an underestimate. New Zealand has a lot of remote coastline, and strandings may occur that we are not aware of. But given the rate of interference for known sperm whale strandings, we believe that the rate at which this happens is relatively low, and that the majority of ambergris found is likely to be found beach-cast."

* * *

After speaking with Anton van Helden, something had continued to puzzle me. Ambergris obtained directly from a whale carcass is black, viscous, and foul-smelling. "I've seen stuff out of a whale because, you

know, the odd whale does wash in, and the odd person has got it out,"
John Vodanovich had confirmed when I spoke with him. "It's just sort of
black, and it really stinks." Bernard Perrin had told me, his voice filled
with Gallic indignation, that he didn't even consider this material to be
ambergris, and he refused to sell it to his customers. The smooth, white
boulders that Perrin sells to perfumers like Chanel have spent years in
the ocean, slowly transforming into the refined high-grade ambergris
that commands the highest market prices.

If fresh black ambergris is so unpleasant and unrefined, why would
anyone want to remove it from a whale in the first place? Who would have
any use for it? These were questions I was unable to answer. For months, I
remained confused by the historical accounts, which told of whalers who
cut enormous soft boulders of ambergris from whale carcasses and sold
them to perfumers, becoming rich in the process. Finally, hoping to solve
the puzzle, I contacted Charles Sell at Givaudan again.

"This question has also vexed me," Sell replies. "Günter Ohloff showed
very clearly how chemistry was responsible for conversion of the raw
whale secretion to the highly prized gray ambergris, so the dark brown
sticky stuff they would have taken from the whale's insides would not be
the same. I have a lump of fairly fresh ambergris, and it smells mostly of
scatole."

Present at high concentrations in feces, scatole is the organic compound that is primarily responsible for the characteristically pungent
unpleasant smell of fresh dung. It is also abundant in rotting flesh. In
chemical structure, as is the case with so many powerfully malodorous
compounds, its structure provides no clue that it will smell completely
repulsive. It consists of two adjacent, compact-looking carbon rings—a
benzene ring and a pyrrole ring—which are known in combination as
an indole ring.

Luca Turin wrote in *The Secret of Scent: Adventures in Perfume and the
Science of Smell*, "Indole, probably the most unfairly maligned molecule
on earth, smells bitter and inky, but is an essential component of all
raspy-voiced white flowers like lilies, tuberoses, etc." This is true. Several species of flower—freesias, lilies, and narcissus, for example—derive
their heavy, almost narcotic fragrances from the presence of indoles at
fairly low concentrations: a familiar cloying, indolic sweetness, reminis-
cent of wilting flowers, of decay and entropy. But scatole is like indole on
steroids. Abrasive and discordant, at higher concentrations scatole is a
rumbling, sharp-edged base note that threatens to overpower everything
else. When diluted, though, it imparts a lasting warm and woody tone to

fragrances. Either way, the odor profile of scatole is a powerful one. In 1948, when Parisian perfumer Victor Hasslauer observed that ambergris had an indolent note, he was referring to the unmistakable presence of scatole, abundant and overpowering in lumps of fresh ambergris like the one Sell is describing.

Sell and I both continue to wonder why fresh scatolic ambergris from a whale would be of any use to perfumers, or to anyone else for that matter. Eventually, Adrienne Beuse provides me with an answer. Not all the ambergris found inside a whale, she tells me, is necessarily of the same grade. In 1891, when Louis Smith crawled the length of a whale carcass to obtain the Bank Lot, brokers in London discovered that its core—a five-pound lump, shaped like a rifle bullet—was a much smaller sample of the finest quality ambergris available. A fine gray core, surrounding by soft layers of reeking ambergris that resembled wet, black clay. Beuse explains:

> Sometimes the only material found inside the whale is this black tar-like liquid, which will be full of beaks. You could say that it's the precursor to ambergris. It's very thick and dark and totally disgusting, you know. But, in amongst it sometimes, lumps– harder lumps of ambergris—get found, and this is what people go searching for. These hard lumps do have a value, and they seem to have some curing that has taken place inside the whale. We quite often see it with a dead whale that they'll pass ambergris before they actually die and reach the shore. Sometimes then they'll get ripped apart by sharks and that might also end up releasing it, and somebody will find a hard lump of ambergris in association with a whale stranding. That ambergris will have a value.

It was because of that value that the Sri Lankan villagers clumsily maneuvered a backhoe across the sand in order to take apart the whale, which had stranded on their beach.

"But," says Beuse, "any ambergris that comes directly from a whale is a different kettle of fish completely from something that has been thirty years out in the ocean, you know?"

* * *

Around this time, deciding I have nothing to lose, I write to Robert Clarke again. "He is 90 years old," his wife had written several months earlier, "and has a more or less serious disease." Whatever his ailment, I hoped it was of the less serious kind. To my surprise, Clarke responds, and we begin a courteous and slightly lopsided e-mail correspondence. On several occasions, I send him a long list of questions I hope he'll be willing to answer. Usually, he writes back quickly, but sometimes it takes two or three weeks, and then occasionally he doesn't respond at all. "I should mention that I am ninety years old, and a bit like ambergris myself," he explains.

I request a photograph, which he sends a few days later: a wizened old man, sitting at his desk, a bookshelf filled with journals behind him. Mostly bald now, he stares blue-eyed and intent, directly into the camera. Two scholarly tufts of snow-white hair reach his ears. He wears a bow tie and a pinstripe jacket, with a white handkerchief neatly folded in his breast pocket. On his desk, at his bent elbow, stand two large cylindrical glass jars. On their yellowing handwritten labels, one word is clearly legible: AMBERGRIS.

I write to ask him about the day in December 1953, when the enormous boulder weighing 926 pounds was harvested from a sperm whale on the deck of the *Southern Harvester*. A photograph of the huge blunt-ended boulder, swinging from a block and tackle like a misshapen wrecking ball, had been included in Clarke's "The Origin of Ambergris," a copy of which still sits on my desk. And in the photograph that Clarke has sent me, one of the jars contains two pieces taken from the same large boulder. They sit like two lumps of coal, black and glittering mysteriously, at the bottom of the dusty jar. The label on the jar reads: "Pieces heavy with CRYSTALS from the external layers."

Clarke writes back that he was not aboard the *Southern Harvester* the day the boulder was taken from the whale. Instead, months later, he traveled to the headquarters of the Christian Salvesen and Co., in Leith, near Edinburgh, and had inspected it there.

"However, I have a story to tell you about the 160 kg boulder shown in Fig. 2," he writes, referring to another grainy black-and-white photograph from his paper, of a large misshapen mass of ambergris, resting on the ship's deck and surrounded by a ring of curious whalers.

In 1947, I was Whale Fishery Inspector on board the *Southern Harvester* in the Antarctic. Late one afternoon, work was finishing on the main deck, and a wire hawser was about to sweep the guts of a whale overboard, when I saw a swelling in the large intestine. "Stop—amba [ambergris]," I cried. The gut was not two feet from the ship's side. Cutting away the intestine revealed the boulder of ambergris weighing 155 kg shown in Fig. 2 of my paper. On page 19 of my paper, I mention that the Government Chemist in London began in 1957 an elaborate analysis of sperm whale faeces, but I have heard nothing of the results, nor of the analyses of ambergris also done at this time.

More than fifty years later, Clarke is still waiting.

*   *   *

"The advances of synthetic chemistry in recent years have not only made it possible for chemists to imitate exactly the composition of the compound," reported a 1928 article on ambergris in the *Sydney Mail*, "but also to produce artificially other and better aromas at one-hundredth part of the cost of ambergris."

The future had arrived. By the 1930s, organic chemists like Max Stoll and Günter Ohloff at Firmenich were constructing synthetic ambergris compounds. And by 1950, they mostly understood the chemistry behind its properties. Within a few decades, ambergris had become obsolete: better living through chemistry. At the turn of the twentieth century, a sharp-eyed beachcomber could find and sell a large piece of ambergris, and his family would enjoy a new life with the proceeds. Suddenly, it had become a greasy, strange-smelling burden.

The *Mail* continued:

> Were one to pick up a quantity of the stuff on one of the beaches around Sydney, where in the past some fair-sized lumps of it have been found, it is even doubtful whether it could be disposed of locally with ease. Some years ago several fishermen found a piece on a beach in the vicinity of Catherine Hill Bay, north of Sydney, and although it was genuine ambergris they hawked it around the city for some weeks before they eventually chanced upon a buyer, and the offer which was made for it quickly dissipated their dreams of a prosperous and easy future.

Today the ambergris market has never been busier. Once again, demand exceeds supply. Ambergris now sells for approximately $1,000 per pound. The very best white ambergris is almost priceless. Why the resurgence now? First, there are limits to molecular chemistry. A substance as complex and singular as ambergris simply cannot be synthesized. And its journey, from sperm whale hindgut to remote coastline—separated by a decades-long interval spent in the ocean—cannot be replicated in a test tube.

We live in an era of specialization. People adhere to strict and demanding diets: veganism, fruitarianism, and raw foodism. There are gardening enthusiasts who choose to grow only heirloom tomatoes, blush pink and shiny on their vines, or enormous oversize pumpkins. A friend of mine has become an apiarist. Another spends every Sunday afternoon dancing, but exclusively to medieval music. We are all experts in something. Inevitably, someone will want to use genuine ambergris. The synthetic versions, they will tell you, lack an authenticity that can only be obtained from years spent adrift in the ocean. Through the Internet, anyone with sufficient funds can obtain the highest quality white ambergris from online-based vendors in New Zealand, France, Italy, or Taiwan.

When researching ambergris one night, I register with an online business-to-business website called Trade Boss, which allows importers and exporters to connect with one another to trade goods. There, traders sell everything from emu oil, pork gallbladder, and guano, to goat hair, snake venom, and ox bile powder. Within a week, I begin to receive automated e-mail updates informing me of new sellers and products. One day I receive an offer to buy a baby capuchin monkey from South Africa. Another time I see a request from Poland for scorpion venom.

I have inadvertently stepped into another world. It is a virtual market, where ambergris sellers in Indonesia sell blocks of ambergris worth hundreds of thousands of dollars to buyers in France or Dubai. In this strange and unfamiliar world, deals take place in luxury hotel rooms in Singapore and Kuala Lumpur. It is the sort of alternative reality in which diplomats from the Middle East might arrive in Australia, pack five or six empty suitcases with ambergris, and simply disappear.

Suddenly, I am receiving regular messages from ambergris traders all over the world. One of them is Peter Chiu in Taiwan. I ask Chiu via e-mail if he will sell me a couple of grams of ambergris. He responds that he deals instead in kilograms of ambergris. When I pose as a perfumer and tell Chiu I am trying to find a source for high-quality ambergris, he offers to send me two grams for free.

For several months, I trade e-mails with a Ukrainian dealer calling himself Valentyn. One day he explains in broken English that because of the restrictions on importing ambergris to the United States, I will have to fly to Odessa to buy the pound and a half of ambergris he is selling. I have no intention of buying his ambergris and don't even know how to respond.

"Greetings!" he writes a week later. "Well that you have decided to buy ambergris? I to you could and is cheaper sell. You would arrive to us on Ukraine we will tell the econom class and would buy all personally from

Tom Donaghy and Geraldine Malloy of Wellington break up a large block of suspected ambergris on Breaker Bay, in September 2008. Credit: *New Zealand Herald*.

Dr. Robert Clarke in his home in Pisco, Peru. One bottle contains pieces of ambergris from the 926-pound boulder taken *Fl. F. Southern Harvester* on July 23, 1953, from a sperm whale in 53°21′S 14°13′W. The other bottle contains a single ambergris concretion of 1 pound 15 ounces from a sperm whale examined in Iquique, Chile, on August 31, 1960. Credit: Aravec Clarke.

Seven-year-old Ben Marsh holding the ambergris he swam into at Oakura, near new Plymouth, New Zealand, in March 2009. Credit: *Taranaki Daily News*.

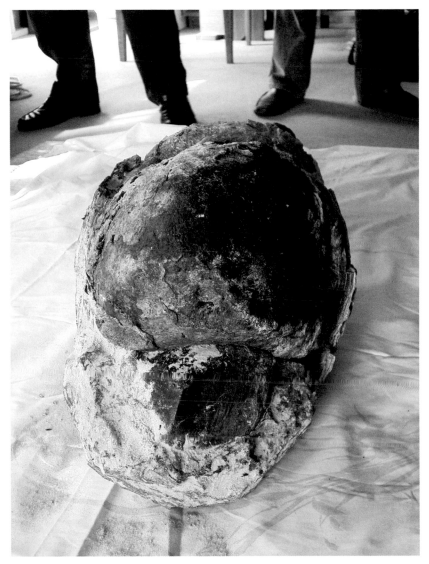

A 200-pound boulder of ambergris, which washed ashore in New Caledonia in 2007.
Credit: Bernard Perrin.

Loralee Wright, with a piece of ambergris weighing thirty-two pounds, which she and her husband, Leon, found on Streaky Bay in South Australia in 2006. Credit: Loralee Wright.

Mike Hilton in his office with several pieces of Stewart Island ambergris. Credit: Christopher Kemp.

Coming in at low tide to land on Doughboy Bay, on the remote west coast of Stewart Island. Credit: Christopher Kemp.

A piece of ambergris, in Department of Conservation ranger Simon Taylor's hand, that had washed ashore on Doughboy Bay. Credit: Christopher Kemp.

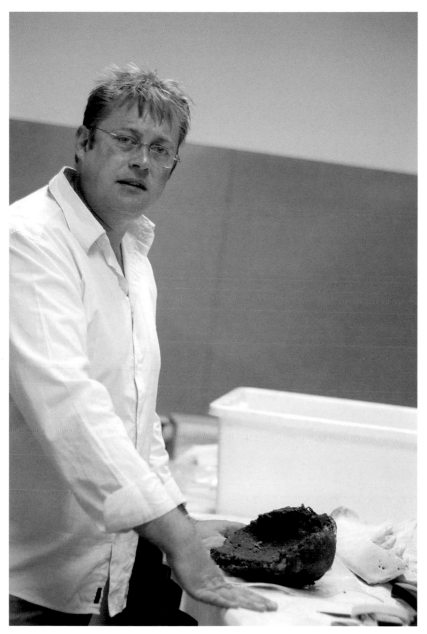

Anton van Helden, marine mammals collection manager at the Museum of New Zealand Te Papa Tongarewa, with a large piece of ambergris in the museum collection. Credit: Christopher Kemp.

A piece of ambergris weighing almost two and a half pounds in the Auckland War Memorial Museum collection, found at Ruapuke Beach, near Raglan, in February 1992. Credit: Christopher Kemp.

Ambergris collected by J. Henry Blake and donated to the Department of Mammalogy, Harvard Museum of Comparative Zoology. Credit: Christopher Kemp.

A piece of Adrienne Beuse's ambergris. Credit: Christopher Kemp.

me. It would exclude any deceit for my part. Money for operation to the
father are necessary to me, I cannot long bargain."

I'm not sure what Valentyn means. But I know I'm not going to Odessa.

* * *

Perhaps the only error worse than finding sewer grease and believing it
is ambergris is finding ambergris and believing it is sewer grease. On a
warm clear night in the summer of 1899, in the squid-filled waters near
Nova Scotia, the crew of a ship called the *Squantum* did just that. A few
of the men were attempting to pass a rope around a large floating mass
as it bumped and slid along the starboard side of the ship. "A torch was
lighted and further and closer examination disclosed that the find was
a mass of greasy or rather waxy substance," reported the *New York Times*.
"It was grayish in color, with here and there a streak of black, giving it a
marbelized appearance."

Someone suggested launching a dory to collect the lump and bring
it back to the ship, but the crew decided against it. A crew member suggested waking Bill Griffin, the ship's captain, who was sleeping soundly
in his quarters. "But this met with no approval. It was bad enough to call
him when there was something in the wind, but to get him out for what
might be nothing at all would be dangerous. For as likely as not he might
be dreaming of roast sparerib and apple sauce or fish at $7 a quintal, and
it would be wrong to spoil his dreams upon uncertainties."

Instead, anxious crew members hung over the railing, peering into
the dark water. They had begun to prod the floating lump with gaffs, hoping to haul it aboard. It was, one of the sailors later said, like poking a
pumpkin with a pitchfork. The crew watched the lump rolling around
in the water. "It became rather tiresome holding on to the stuff with the
tide pulling hard all the time and an hour or two of sleep already lost,"
reported the *Times*. "So it was decided to let it go adrift and consider the
incident closed."

And the lump floated away on the black water. On deck, Moses Bridges
quickly pocketed a small piece that he had managed to carve away from
the boulder. He kept it in a tin. Several times during the remainder of the
journey, Bridges took the greasy little nugget from the tin and greased his
boots with it. When the *Squantum* arrived in Boston and Bridges had the
object assessed, the crew discovered it had discarded an enormous boulder of ambergris.

Again and again, I found cautionary tales just like this one. In 1903, somewhere off the coast of Brazil, the crew of the *Kelvinbank* noticed what the *Washington Post* reported was "a spongy mass of some greasy substance. It was black, with the exception of a few places, where it was mottled like marble." Two of the crew dropped a rope over it and hauled the lump aboard. None of them had seen anything like it before. Days later, left to sit on the deck, it began to smell. After using some of it to grease their boots, the crew threw most of it overboard. The remainder was thrown in the slush bucket, along with the refuse from the galley. On arriving in New York, the sailors learned they had discarded approximately two hundred pounds of ambergris. Five years later, in the summer of 1908, a vessel called the *Antiope* arrived in San Francisco after a rough two-month voyage from Australia. A few days after leaving Newcastle, about 150 miles north of New Zealand, she had encountered a powerful storm, which had swept her cargo from the deck and ripped the sails from her masts. When the ship finally limped into harbor, the British crew had in its possession a few handfuls of a strange greasy substance. It had been collected at sea. It was all that remained of more than fifty pounds of the material, which the crew had used to grease down the *Antiope*'s masts. Taking what small amount was left, the men greased their oilskins and weatherproofed their boots. From the few handfuls that remained, an Oakland druggist confirmed that it was genuine ambergris. Reading each of these accounts, I was determined not to make the same mistake.

*   *   *

I am standing on the curving beach at Aramoana trying—as I can only imagine Sir Francis Godolphin did along the Cornwall coastline—to covet the wind as it blows in briskly from the sea.

Aramoana: in the Maori language it means "pathway to the sea." It is a beautiful, remote, and windswept place—a wide belt of sand that sits near the narrow mouth of Otago Harbor. Across the water, on the tip of the peninsula, is Taiaroa Head, where a simple lighthouse perches on the towering cliffs. I have walked along the grassy edge of the cliffs there and looked over into the dark wet space below, which is filled with rocks the size of apartment blocks, ringed by an impenetrable tangle of bull kelp. Waves boom down in the hole. After each wave, the high-pitched sound of retreating water, a million droplets falling through the rocks as they make their way back to the gathering sea. Twenty-foot swells. A muscular ocean. A tower of spray hangs in the hazy air. It is surprising each time and impossible not to watch. The bull kelp below is as thick and rubbery as flattened car tires. Each flat green blade on every long thick stem straightens as the water retreats, and then corkscrews and collapses into wet folds against the rocks as the waves heave forward again.

I had come here months earlier with my wife to collect edible seaweed. We stood on the rocks hauling in slippery armfuls of bladder kelp and wakame, harvesting the rocky pools for frilly purple-green karengo. We collected sea lettuce, which grows in thin sheets like transparent green cellophane, and sprigs of Neptune's Necklace and Dead Man's Fingers. Standing on the shoreline, we bit into them, tasting the seawater that dripped from their stalks. Back home, I'd removed the air-filled bladders from wet lengths of bladder kelp and pickled them in jars. We filled our basement with clothes racks and draped wet strips of seaweed over them to dry.

Over time, I had learned to enjoy the smell of the mudflats at low tide and the briny tang of dried seaweed. Eventually the smell infiltrated every corner of our tiny house. But that was in the spring. The skies were blue then. It is autumn now. The ocean looks hard and cold. On the horizon, the water and the sky are indivisible. Gulls stand near the surf, battered into stillness by the wind. I lower myself toward the wet sand like a sprinter planting his feet in the blocks. I raise my head and straighten my spine, which allows me to look along the shoreline, curving away to the east. And then, rather self-consciously at first, I begin to covet the wind between me and the places I search.

For a long time, I have been convinced that these clean and windswept beaches could act as a catchment area for ambergris. Before the tidewaters enter the harbor, surging between the heads to the east, all kinds of marine debris must lose its momentum, finally coming to rest and washing ashore on the beaches near the mouth of the harbor. Ambergris, floating in the ocean for decades, has to end its journey somewhere. Why not here?

In the distance, a lone dog walker throws a stick into the surf for a wet dog to retrieve. A family plays in the sand nearby, determined to ignore the cold wind that bends the dune grass over until the tip of each stalk touches the sand. Undaunted, I squat near the tide line, fox-like with my nose held in the stiff tidal breeze, trying to sample and discern the smell of every molecule as it passes my olfactory receptors.

"It is pretended that foxes, according to Frèdol, are very partial to ambergris," wrote Armand Landrin in 1875, in *The Monsters of the Deep: And*

Curiosities of Ocean Life, "and that they resort to the seashore in search of it. They eat it, and restore it much in the same condition as when they swallowed it, if we refer to its perfume only, though greatly altered in colour. To this cause is attributed the existence of some fragments of whitish ambergris in the Aquitanian Landes. Locally, they are called *ambre renarde*, or *fox's ambergris*."

I doubt this, but I don't have time to think about it. The winds are growing stronger. The sky is beginning to fill with dangerous-looking rain clouds that roll across the sky, swallowing other clouds, until a black billowing front has completely enclosed the bay.

Closing my eyes, I hold my nose aloft in the cold, sand-filled gusts of wind sweeping in from the ocean, and I proceed slowly along the high-tide line sideways like a crab, in an uncomfortable half-crouched position. Pressing eastward, dark cliffs propping up the pewter sky in front of me, I serve as a counterpoint to their quiet dignity, stumbling around on the sand and dragging behind me two long wet streamers of wakame.

*   *   *

It is almost noon when I call John Vodanovich in Dargaville. In the weeks since we last spoke, I have thought of the ambergris collector often, wondering if his efforts to find ambergris have been more successful than mine.

"Johnny's on his way down now to have a look," says Kim Soole, Vodanovich's partner. "He's already been down this morning. The sea has been quite high; the winds have been good." In other words, the weather has turned again in their favor. It is a good time to search for ambergris on the long windswept beaches of Northland. For the next few days, they will scour the beach after every high tide, in heavy rains and battering winds if necessary, in the hope that the rough seas have dumped ambergris on the shore. "He went down yesterday and found a bit," she continues, "and he found a little bit this morning. It was high tide this morning at eleven thirty-eight, so we usually go down after the high tide."

Vodanovich and Soole have a secret weapon in the hunt for ambergris: a dog's sensitive and tireless nose. "Dogs will be able to find a pinhead," she says. "They dig it out. John found a bit in a bank the other day and it was obviously buried in the bank, but the dogs sniffed it and dug it out."

For a moment, Soole's voice is muffled as she asks Vodanovich, "What was it you found yesterday, John?" In the background, I hear Vodanovich's respond in a low-pitched melancholy rumble. "Fifty-something, I think," says Soole, "fifty-nine grams. It's a nice whitish-gray piece too. He went to Ninety Mile Beach the other day and got about 300-odd grams. That's pretty good really because the week before they went to Muriwai, and they got hardly nothing. I mean they always come back with some, you know, particularly when you've got a dog."

*   *   *

On the beach, hobbled by my glistening dark-green leg irons, I finally stop walking. It is cold in the shadows beneath the cliffs. Even the dog walkers and the stubborn families have gone home. I am alone on the sand, with my feet encased in seaweed. I begin to wonder if I should get a dog.

*   *   *

"You've never seen it before, have you?" Mike Hilton asks me when I first knock on his office door. The University of Otago geography lecturer is standing by his desk, holding a small Tupperware container. Inside, he says, is ambergris. The two old black pieces of ambergris I had held a month earlier at the Otago Museum had smelled only of camphor, and since handling them, I had begun to wonder if they were ambergris at all. "Let me show it to you," Hilton says, handing me the container. Resting incongruously on a blue Cookie Monster napkin at the bottom are three small pieces of ambergris.

"I like to lift the lid and have a whiff," he says, and nods encouragingly, "so go ahead." He watches eagerly as I slowly peel up the edge of the lid and place my nose at the opening.

A powerful odor fills my nostrils. It is revelatory: a breakthrough. My brain swims. For a moment, I think I am going to sneeze, fighting a sensation that begins as a tickle in my nose and then spreads, filling my throat, and completely occupying my sinuses. All at once, I smell: old cow dung; the lumps of wet, rotting wood that I have kicked along the beach; tobacco; drying seaweed; and the grassy open spaces of Aramoana and Long Beach. And, beneath it all, something indescribably elemental. It is a mixture of the low and the high. The unavoidable and the unobtainable.

A mutual friend had introduced me to Hilton. Small and solidly built, he has a prominent nose and a crest of thick white hair on top of his head. He reminds me, not unkindly, of an inquisitive cockatoo. Bending toward the ambergris on his desk and breathing in deeply, I expect his white crest to rise in appreciation. These black-and-brown nondescript fragments, he tells me quietly, came from Stewart Island, a rugged and isolated island located off the southern coast of New Zealand. For several weeks a year, Hilton is based on Stewart Island, collecting the data necessary for his fieldwork. He tells me they were all found by a graduate student who sometimes accompanies him on his fieldwork, near the high-tide line on Mason Bay, a remote stretch of coastline located on Stewart Island's exposed west coast.

"She has a nose for it," he says, a little downcast. "I've never found any."

And then he bends again birdlike toward the ambergris.

It is late afternoon. Hilton sits at his desk now, carefully balancing a piece of the ambergris on a pair of scissors. His forehead is furrowed with concentration. He is holding the closed blades in his hand, and the ambergris sits like a flat black stone, canted precariously in the oval-shaped hole designed to house his thumb. Moments earlier Hilton had been talking. But his voice has trailed away, slowly at first, now stopping altogether. I am holding my breath. One by one, he uses the handles of the scissors to transfer the three small pieces of ambergris from the plastic container to the cover of a textbook, which he then lays reverently on the table in front of me. It is a watershed moment. Finally, I am sitting in front of three genuine, pungently aromatic pieces of ambergris. I can smell them from several feet away.

In fact, I am permitted only to smell them from a distance. The oils on my hands, Hilton says gently, will damage their pale weathered surfaces—hence the scissors and the balancing act.

The three small and irregularly shaped pieces of ambergris sit in front of us now, arranged in a row from largest to smallest They remind me of mysterious, patinated museum relics. But they are ambergris.

Marbled and pocked, their uneven surfaces are mottled with yellow, green, and white patches like moldy pieces of cheese. In fact, placed side by side, Hilton's pieces of ambergris represent three distinct stages of maturation. The largest is a flat black piece that would fit comfortably in the palm of my hand if Hilton would allow me to hold it. The other two pieces are progressively smaller and whiter—the smallest is a little white nugget, the size of my thumbnail, white and smooth, with a waxy resinous exterior. They all look like pebbles that, at other times, I have walked past and ignored on the beach.

I ask Hilton to characterize its odor. "It's musty," he says, after inhaling and pausing thoughtfully for a moment. "Do I get a hint of fish in it?" he asks, raising his head to look at me before lowering his nose again.

Mike Hilton smelling his ambergris. Credit: Christopher Kemp.

"There's a hint of rotting wood," he says. "It's a new smell to me. It's like nothing I've smelled before."

The odor forms a nimbus around my head and takes up residence in my nose. When I close my eyes, I am transported to a horse stable on a warm afternoon, when the heat of the sun has warmed the hay, filling the air with the smell of livestock and the sweet aroma of the wooden joists crisscrossing the stable roof. For a moment, I am there—and not here, in this small clean office, overlooking the modern utilitarian campus buildings of the University of Otago campus.

On several different occasions during his frequent fieldwork trips to Stewart Island, Hilton tells me that he has encountered professional ambergris collectors walking the isolated west coast beaches. The most intriguing of these is a shadowy figure known as Fisherman Phil. "He's a fisherman," says Hilton. "I only know him as Fisherman Phil. This is his off-season, and he spends it searching for ambergris."

The last time Hilton had seen Fisherman Phil, the ambergris collector was on Doughboy Bay, a remote and starkly unpeopled west coast bay that is simply miles from the nearest civilization. "He'd been at Doughboy for six weeks," says Hilton, "and was living in a cave. He and his wife. He had this cave decked out. Every day—the bay is a curve of three kilometers— he walks the high-tide line, and then he turns over every log, every bit of debris in the bay. So, he really sieves that bay. He has some oceanographic theories about storms that turn up ambergris on the coast."

We pause to smell the ambergris again. When I think of Fisherman Phil, I am reminded of the lengths to which people will go, walking miles twice a day and spending hours lifting heavy wet pieces of tide-swept lumber, to find ambergris. "So every piece of timber that he could lift had been lifted and looked under," Hilton continues. "When he gets bored of Doughboy Bay, he walks over the hill to Mason Bay. He did say that this year was a bad year, but I think he might have said that about the last couple too."

As I sit in Hilton's office, I decide to visit Stewart Island and try to find Fisherman Phil. Hilton prints out maps of Stewart Island for me, circling Doughboy Bay and Mason Bay on the west coast. He carefully arranges the pieces of ambergris on top of the textbook, moving them into place with the tip of a ballpoint pen. I take photographs of them. During my visit, ambergris fumes have diffused, filling Hilton's small office with an indescribably complex aroma of dung, old wood, and the sea. When I leave, the smell of ambergris stays with me for hours afterward. Somehow, I am still able to smell it when I climb into bed that night. Only when I wake up the following morning do I realize I'm unable to smell it anymore.

And I miss it immediately.

7 THE HOPEFULS

*What is probably the largest and most valuable quantity of ambergris secured for
many years was discovered the other day at Doughboy Bay, a few miles south of
Mason's, by Mr. Charles Yunge, owner of the fishing launch Scout, with two com-
panions, Messrs. B. Bailey and J. White. The find weighed 16 lb. in all, one lump
tipping the beam at 8 lb. ✳ New Zealand Evening Post (November 5, 1924)*

*A piece of ambergris weighing approximately 60 pounds, the largest ever
found in the Foveaux Strait area, has been placed in the custody of an Invercargill bank. Its value is estimated at well over four figures. This valuable find
was made recently by Mr. F. Traill, Half Moon Bay, while patrolling beaches on
the west coast of Stewart Island. ✳ New Zealand Evening Post (July 4, 1945)*

"I think we'll just come in and land on this one," the pilot yells to me over
the drone of the engine. "We'll take a shit-kicking if we try to come back
around. Just stay down."

We're flying at low altitude in a small single-engine Cessna, over
the remote and bush-clad west coast of Stewart Island. The ground beneath us is like a map—a series of long sandy bays and rocky headlands
stretches to the south, punctuated by coastal inlets that become rivers,
looping through the hilly, overgrown interior. And everywhere else, the
deep blue of the ocean. It is early in the morning. Half of the forest is
still in shadow. Strong easterly gusts of wind rock us, suddenly pushing
us sideways. We pitch forward in our seats, lurching first to one side and
then the other, drifting vertiginously over the sea. Before we had taken
off, the pilot told me he would be flying in low over the sand, barely clearing it, so that he could check his approach. And then he would climb
again, coming around a second time to land on the beach. Not anymore.
It is too windy. We hurtle through the air.

I look at the dials and gauges on the cockpit panel. They are mean-
ingless to me, but I watch them intently. A red light is winking. Colliding with another strong gust of wind, we move through the air diago-
nally, leading with the tip of one wing. A rolling green carpet of lush
native forest unfurls beneath us. Through breaks in the forest canopy, I
can see groves of towering tree ferns in the sun. From above, they look
like green bombs bursting. I watch the shadow of the plane climbing
the hills below, keeping pace with us, crossing rocky riverbeds, falling
into dark valleys, and reappearing in the hazy sunlight. The pilot leans

toward me, his head almost touching my shoulder, cranking something near his feet noisily like a wrench. To my right, the sea looks calm and wrinkled with whitecaps. But I know it is cold. Suddenly, I notice the forest has stopped rolling beneath us. We have flown over the green lip of a cliff face. A biscuit-colored apron of sand through the cockpit window, quickly growing larger. Low tide. Grassy dunes to the left. Sea to the right. A final plunge through air, and we land. A moment later, I am standing on Doughboy Bay. The pilot makes a tight turn on the beach and takes off again, droning into the distance. I watch the plane as it becomes a white speck in a blue sky. Then it is gone. I can no longer tell it from the seagulls. I am alone.

In fact, I am slightly more nervous knowing that I am not completely alone: a minute or two earlier, as the Cessna swung and seesawed crazily above the sand, I had seen a lone figure, dressed in overalls, standing on the beach, looking up at us. And now I am alone with that person, whomever it is.

Hemmed in at each end by rocky impassable cliffs and backed by steep overgrown hills, Doughboy Bay is a seriously isolated place. The headlands at either end of the bay almost overlap each other, making it difficult from the beach to catch more than a glimpse of the open sea. To the north is Mason Bay, a challenging ten-hour hike through native forest. Oban is the only town on the island, and it sits on the opposite coast, through twenty-five miles of tangled and untended bush.

I take off my shoes and walk toward the waves. The water is clear and so cold that it numbs my toes. It's still early, but the sun is already high in the sky, and I have to shield my eyes to see anything. Walking back to the circle left in the sand by the departing airplane, I begin to follow the wheel tracks that extend from it. Two straight dark lines in the sand. I walk between them along the beach until they disappear. And I see no one.

* * *

I had come to Stewart Island for five days, boarding an airplane in Invercargill on New Zealand's South Island for the fifteen-minute flight across the treacherous waters of Foveaux Strait. While I was here to find Fisherman Phil and locate the cave on Doughboy Bay in which he spends weeks every winter with his wife, I wanted to find a piece of ambergris for myself. I had brought with me *Rakiura* (1940), the definitive history of the island by Basil Howard, and *The Stewart Islanders* (1970) and *In the*

Grip of an Island (1982) by Olga Sansom, who was born and raised on Stewart Island. From these and other older texts, I had come to understand that Stewart Island has always had a connection with ambergris like nowhere else on Earth. In a sense, I was following Mike Hilton's ambergris home too. This, at least, was where its journey had ended: on the high-tide line, near the vegetation on Mason Bay—a remote twelve-mile long crescent of sand that stretches along Stewart Island's rugged and exposed west coast. Where its journey had begun, the path it had taken, or even how long it had been floating in the sea before Hilton's graduate student found it were simply unknown, and mostly unknowable. Even the most reliable and up-to-date computer modeling software cannot tease apart mysteries as complex as these. But its journey had ended here.

If New Zealand is shaped like the letter *i*—a listing green *i* adrift in the southern seas—then Stewart Island is a tiny misshapen dot that sits beneath the canted upright: an overgrown, hilly, and half-forgotten dot in a vast and endless ocean. The Maoris settled here first—as early at the thirteenth century—naming the island Rakiura, which means "Land of the Glowing Skies."

In March 1770 James Cook sailed the HMS *Endeavour* along the south coast of Stewart Island. He was the first European to do so, narrowly avoiding shipwreck against a long rocky ledge he named The Traps. Rounding the rocky southern coast, Cook concluded that Stewart Island was not an island at all, but a peninsula. Referring to his own sketches, the overlapping outlines of the numerous small islands in Foveaux Strait had deceived him. Accordingly, he named the island South Cape. In early maps drawn from expedition notes, a dotted line tentatively connects the island to the mainland. On the day the *Endeavour* navigated the southernmost part of New Zealand, Cook celebrated a junior officer's birthday by ordering that a dog be killed: its hindquarters were roasted; a forequarters went into a pie; and its stomach was used to make a haggis for the Scotsmen aboard. Thirty-four years later, an American sealer named Owen Folger Smith finally discovered Foveaux Strait and navigated its narrowest parts in a whaleboat. He charted the channel, naming it Smith's Strait. And then in 1809, William W. Stewart, the first officer of the Australian ship *Pegasus*, determined the island's northernmost points. The island officially became an island and was marked as such on maps. It was named in Stewart's honor.

In *Rakiura*, Howard described Stewart Island as "irregularly triquetrous in outline." From above, it is shaped like an arrow with a broken tip that points absolutely nowhere. Today around 380 permanent residents call Stewart Island home, a number that has remained stable for several decades. There are a few prominent families—whose numbers are slowly dwindling—and a quiet and steady influx of outsiders to replace them. Almost all of them live in Oban, which is now the only settlement on the island. Strung around Halfmoon Bay, on the island's sheltered east coast, Oban is little more than a few stores and homes clustered along a winding coastal road, from which a handful of named streets radiate. Over on the west coast, a series of isolated beaches: Smoky Beach, Ruggedy Beach, Big Hellfire and Little Hellfire beaches, Mason Bay, and Doughboy Bay. It is challenging terrain—mile after mile of exposed and uninhabited coastline. A few weeks earlier, I had arranged to fly from Oban to Doughboy Bay. The pilot would land on the beach at low tide and leave me there, returning twelve hours later during the following low tide, to pick me up.

In 2002, 85 percent of Stewart Island was designated Rakiura National Park and is now protected. In other words, almost all of Stewart Island's 675 square miles are unpopulated. Most of its residents live within the couple of square miles that contain Oban and Halfmoon Bay. By contrast, Singapore is less than half the size of Stewart Island, with a land area of 274 square miles. It has an estimated population of almost five million inhabitants. Elsewhere on Stewart Island, abandoned villages rot into the ground, forgotten: Port Pegasus to the southwest, a booming tin-mining town in the 1890s; the Maori settlements at Port Adventure to the southeast and at Maori Beach on the north coast; all dismantled by the wind and picked over by the gulls a long time ago.

But Oban remains, clinging tenuously to life through commercial fishery and tourism—and stubbornness. It is a small place, in every sense. The Stewart Island telephone directory is a single-sided, laminated piece of paper. In total, there are fewer than three hundred phone numbers on it: plenty of Johnsons, Leasks, and Squires; a number for Conner A (Squirt); and Big Glory Seafoods. There is a general store, but no banks or fast-food outlets. Residents only make their way to the post office after they have heard, passing overhead, the familiar drone of the airplane that delivers their mail. There are cars, but almost nowhere to drive them. There is one school on the island—Halfmoon Bay School, which sits a block or so from the wide and spacious harbor. This year, nineteen students will attend classes there. Older children attend high school classes "in town," which is how Stewart Islanders refer to Invercargill, the southernmost city on the mainland. Town is also where many Stewart Islanders get their hair cut, boarding a ferry from the wharf in Halfmoon Bay, or taking a flight across the 20-mile-wide strait, and stopping off for a quick trim while they stock up on supplies not available on the island. Several times a year, a hairdresser makes it over to Oban from the mainland, converting the fire station into a makeshift salon and cutting the hair of anyone who needs it.

*   *   *

Back on Doughboy Bay, it is midmorning and the tide is coming in. I'm still optimistic that I will find some ambergris eventually if I walk the beach all day. And so I begin to scour the margins in the eerie silence, past half-buried brightly colored plastic containers and bales of old tangled fishing nets. I pick up an empty Nikka Whisky bottle from Japan and a dented plastic drink bottle with Korean characters printed on its side, and I carry them with me for a while along the beach. The tide is dumping a colorful drift of flotsam from across the world in the bay: shipping pallets, waterlogged coconut shells, an intact lightbulb, clumps of wet kelp and sea lettuce, a half-eaten apple, multicolored plastic widgets, crumbling Styrofoam boards, empty mussel shells, broken plastic fishing buoys, dead gulls, a pink baby's sock, a long tangled zipper, ripped from a suitcase, and bleached white piles of driftwood that look, from a distance, like antlers rising from the sand. I look for ambergris among the flotsam. More than any other beach I have visited, the sand at Doughboy Bay is littered with pumice. There are pieces of it everywhere. Some are as small as a peanut, but others are large and round, like pitted white cannonballs. After just half an hour bending and straightening in the sun, I have grown tired of picking them up, smelling them, and throwing them aside. At the northern end of the bay, I find the Department of Conservation hut, a small ramshackle outpost used by hikers, hunters, and DoC rangers. Following Mike Hilton's instructions for finding Fisherman Phil's cave, I slowly make my way south from the hut, through the scrubland and along the thick green edge of the forest. A loud screech rises from somewhere within the dense jumble of trees. Craning my neck nervously to peer through the emerald crosshatch of shadows and trunks, I suspect either a lost penguin, a kaka—a species of large green parrot common on Stewart Island—or a heavily armed hunter who has spent too long in the bush.

Suddenly, I see a flash of color in the dunes: two blue dome tents, nestled in the shadows at the base of the hills. I can see someone walking toward me through the grass. Theo Chapman is a biodiversity ranger for the Department of Conservation. I ask him what he is doing on Doughboy Bay. He takes a few more steps toward the tide line and stops. "This is marram grass," he says, pointing to the grass that covers the dunes, growing in bright green clumps between the shore and the forested hills. "It's very invasive," he says. Chapman and his fellow rangers are part of a widespread DoC program to eradicate introduced species like marram grass from the west coast dune systems. He turns and points over his shoulder to a pale circle of dry and hollowed stumps, like straws in the sand. "See these rhizomes?" he says. "These are all that's left from previous treatments."

Marram grass was introduced to coastal areas by the Ministry of Land more than a hundred years ago. It was planted to combat drifting sands and land erosion—caused by the European immigrants who used to set fire to native grasses growing on the dunes—and quickly replaced indigenous species like pingao grass. The practice was continued into the 1980s, but the Department of Conservation is now attempting to reintroduce native species to long windswept stretches of the west coast.

I tell Chapman that I am on Doughboy Bay to look for ambergris and to find Fisherman Phil and the cave where he spends his winters. A moment later, Chapman has matter-of-factly offered to show me two of the three things I'm trying to find. He walks toward the bush, parts a screen of broad shiny leaves with his forearms, takes a few big steps over a bed of ferns, and disappears into the shadows. We walk up a gentle slope in half-darkness and arrive at a cave opening: Fisherman Phil's cave. It is quiet and dark. Although we've only walked fifteen paces or so from the dunes, I can no longer hear the sea. We have stepped into near silence. I can hear only water falling, dripping occasionally from the rocks and landing in puddles on the ground.

In the shadows of the cave, DoC biodiversity ranger Simon Taylor is preparing for another day on the dunes. He stands beneath the rocky overhang, stretching. I tell him that Fisherman Phil spends his winters here in the cave. "I've met him," says Taylor. "And his wife," he adds. "I don't know what her name was, though. Mustang Sally, maybe?" He laughs to himself as he organizes his kit. "He said he was a fisherman, anyway," Taylor says with a shrug. "He seemed really laid-back."

Chapman and Taylor show me a homemade bed they found in the cave when they arrived. It belongs, they say, to Fisherman Phil. Lopsided, it sits in the vegetation outside the cave now. It is simply constructed— a rack of mostly straight pieces of driftwood nailed together and strung with green netting. It looks uncomfortable. Grass has begun to grow through it. It's hard to believe anyone could survive more than a night on such a bed, but Fisherman Phil rises from it each morning in the winter and walks the length of the bay to overturn driftwood and search beneath it for ambergris. The cave is not very deep—maybe fifteen feet from the entrance to the back wall. On one side of the cave is a gas-powered stove. A drying cloth hangs limply from the washing lines strung between the trees. A fire of paper and trash is burning near the cave entrance, sending a column of dark smoke up into the air. A piece of tarpaulin has been tied above it like a stained sail, to direct to smoke away from the opening. It is working, but the rock wall above the cave is blackened with soot and disappears into the green shadows of the overgrown bush. Near the back wall of the cave, four folding chairs have been arranged side by side. As I quietly survey the scene, a cold droplet of water falls from the branches above the cave and lands in my eye.

"Yeah, we're pretty feral," says Taylor, as I wipe my eye with my sleeve.

"That hot shower at the end of it is pretty nice," agrees Chapman, nodding.

Chapman and Taylor and two other rangers have spent most of the last four months on Doughboy Bay, eradicating marram grass from the dunes. Working for ten days at a time, two of them sleep in tents near the cave among untamed clumps of the grass they are trying to remove, and the other two sleep on bunk beds in the stuffy DoC hut toward the northern end of the bay. After ten days in the dunes, they fly to Oban for four days' rest before returning to the coast—maybe to Doughboy Bay again, but more likely to Mason Bay—to resume spraying for another ten days. Back and forth, and then back again. By the time I met them, they were on the eighth day of another ten-day stint. They had been traveling between Oban and the west coast beaches since November. It was now February. They had spent most of the Christmas holidays on the beach, spraying marram grass. They seemed a little feverish to me. The sun had burned them all various shades of red. They no longer noticed the sand flies that settled on us like a black biting blanket whenever we stood still. They were very eager to talk to me about different types of grass.

Then Taylor pulls out a little nugget the color of wet slate. It sits in his palm like a misshapen nut. Its surface has a waxy sheen to it. He says he's not sure if it's ambergris. "I found it yesterday," he says, "on the high-tide line." He smelled it, he says with a shrug, before he saw it. "I grew up on a farm," he explains, "and the smell reminded me of the shearing sheds."

I take it from him and smell it while Taylor and Chapman look on expectantly with their eyebrows raised, like hopeful prospectors. It is ambergris. This dark, slate-colored little nugget was the reason I had come all the way to Doughboy Bay—the remotest place I had ever been—bouncing and seesawing over an unpeopled coastline in a single-engine Cessna: for this little kernel of ambergris. It was a strange moment in a longer and stranger journey.

"Jono will be back later," says Taylor, quickly pocketing his ambergris, "and he's got two larger pieces."

*   *   *

Stewart Island is twenty-five miles wide and stretches forty-five miles from north to south. Its highest point is Mount Anglem on the north coast, which rises to 3,212 feet above sea level. Parts of the northern half of the island are flat and swampy, but, elsewhere and in general, the terrain is rugged, green, and hilly. From above, the flat areas are irregular in shape, like a slowly spreading green oil spill. Long thin ridges of hilly granite sit between flattened fingers of swampland, which extend and radiate from the basin of Freshwater River, a little west of Oban, and stretch northwest alongside the base of Thomson Ridge, toward the Ruggedy Mountains, and westward to Mason Bay. Geologically, the island is mostly granite. A long fault line of schists, marbles, and quartzites—running in a soft band—divides the island and has slowly eroded over time to form the swampy depression that occupies part of the northern and western interior.

From a 1936 study of the geomorphology of the island: "The climate of Stewart Island is disagreeable: the prevailing winds sweep across the highlands driving with them banks of cloud and showers of rain, but snow seldom falls—in fact the top of Mount Anglem is snow-clad for only a few weeks in mid-winter." In fact, rain is one of the most important commodities on Stewart Island—perhaps even more important than ambergris. Most residents rely on rainwater as their sole source of drinking water, collecting it for long-term storage in large tanks. But during my visit, the sky was almost always a cloudless and uninterrupted blue. The winds that prevent trees from growing on the tops of the grassy exposed west coast ridge lines were absent, and the air was still.

Tim TeAika knows those ridges better than anyone else. For twenty years, he maintained more than a thousand sturdy sheep there, leasing ten thousand rolling acres of pastureland that stretched all the way from the coast at Mason Bay and extending inland eastward as far as Freshwater Flats. TeAika is the last surviving member of a breed of men now gone from Stewart Island—he was the last farmer to hold a lease for a sheep run on Stewart Island. He left in 1986. The farmers are all gone now—those pioneering Bunyanesque men who mustered their stock on the west coast, driving them through the wet tussock-filled hollows and across the steep rain-swept ridges above Mason Bay, building their own spartan homesteads there among the sheep and home schooling their children in the remote wilderness.

TeAika is eighty years old now. His hearing is failing, he tells me loudly over the phone. He lives in Invercargill, on the mainland. "There's a part of the run on the Freshwater Flats," he recalls softly, "but we only farmed around Mason Bay. There was about 5,000-odd acres there on our farm and that's all we used. The rest was too rough."

I ask TeAika if he ever found ambergris on the shore at Mason Bay. "No, no, we didn't," he says slowly. "It wasn't one of my interests really. We did pick up a bit after there was a sperm whale washed ashore and disintegrated over the months and months of just rolling around on the beach." That was back in the late 1960s, he says. After the surf had broken up the carcass and the carrion eaters had picked over the rest, TeAika made his way to the beach to retrieve the ambergris. "I think we got about ten pounds," he recalls. "I don't think it was worth a great deal and, being fairly fresh, I didn't bother trying to sell it. We gave quite a bit of it away to schools who were interested in teaching whale habits. Yeah, ambergris wasn't high priority with me. In fact, it was quite low." Instead, TeAika was busy mustering sheep and hunting deer and opossum.

"There was always the opossums," he says, almost to himself. "That was an annual winter job that I had."

* * *

In a *New Zealand Railways Magazine* article in 1939, Ronald McIntosh wrote: "Practically every Stewart Islander has a piece of ambergris, and the menfolk keep it in their tobacco pouches, claiming that it gives the tobacco a distinctive flavour and aroma." After six months combing the west coast beaches, McIntosh claimed, one successful ambergris hunter was able to retire permanently on a fixed annual income.

"In the old countries ambergris had many uses," wrote Olga Sansom in *The Stewart Islanders*, a history of the European settlement of the island. "At Stewart Island we have used it to plug nailholes, to start a fire, to scent a tobacco pouch; and I keep a nice knob of it in my top drawer among the scarves and gloves and handkerchiefs."

Seated next to me in the waiting room of the Stewart Island Health

Clinic, an old man nurses the ragged and tar-filled cough that only develops after years of dedicated smoking. He selects a boating magazine from a dog-eared well-read pile, clears his throat, and says to me, "Well, I can dream, can't I?" He laughs and starts to wheeze: like a westerly pounding the coast, it begins slowly, deep within his chest, and builds steadily until I feel like I am sitting next to a bellows. The long breezy rattles bend him over in his chair. I look through the stack of magazines. They are all about boats. When the man finally stops coughing, I lean toward him in the sun and quietly ask him if people find much ambergris nearby. "Not here," he says, still flicking through his magazine. "Over on the west coast, on Mason Bay." He leans closer, speaking conspiratorially. "One of the blokes built a boat with it," he says under his breath, "found enough that he sold it and built a boat."

"Who was it?" I ask.

"It was one of the Leasks," he says. "Must have been a bloody big piece."

He begins to cough again, wheezing in the sun like an antique harmonium. "But not here," he continues breathlessly, "it's mostly westerlies, you see." Westerlies bring flotsam to the west coast, which acts like a vast net, but we are sitting in a sunlit waiting room in Oban, on the east coast, road-less hilly miles from the ambergris delivered to the other side of the island.

"Mason Bay, though," the old man says, "is a good few miles of beach."

*   *   *

Long before Fisherman Phil took up his regular winter residence in the cave on Doughboy Bay, it was used by Adam Adamson: the Ambergris King. "In four years," Adamson had told a *Poverty Bay Herald* reporter in July 1914, "I have found, I suppose, from 200 to 300 pounds worth of ambergris washed up by the drift current and buried in the sand." It was winter in New Zealand, and Adamson had made the journey from his homestead on the west coast of Stewart Island, across the Foveaux Strait to Bluff, to sell his ambergris. "This time I brought in one piece weighing 20 oz. alone," he said, "and other odd bits, weighing altogether about 50 oz. I found the big bit on Mason's Bay beach."

Adamson was a rarity: a talkative and successful ambergris hunter. A 20-ounce piece of ambergris is a significant find anywhere. When Ben Marsh found a lump half that size near New Plymouth in March 2009, he needed both hands to hold it. Adamson's single largest piece of ambergris would have been the size of a grapefruit, or one of the larger pieces of pumice I had spent all morning lifting and smelling, and throwing back into the seaweed. Depending on its quality, if I found a piece of ambergris like that on Doughboy Bay, it might be worth $10,000 or more today to a wealthy buyer in Asia or in the Middle East. And three hundred pounds of ambergris—the amount Adamson claimed to have found in four years—would be worth $1.3 million. I had been scouring the bay all morning for even the smallest piece of ambergris, and I had found nothing. After lifting driftwood to look beneath it for the last hour, I would have been satisfied with a little nugget the size of my thumbnail. Some of the larger pieces of driftwood were much heavier than I am: long tapering sun-bleached trunks, riddled with wormholes and covered with the tracks of borers, their branches snapped off by the currents of the ocean. It was hard work in the sun. I soon grew tired and bored of walking up and down the same stretch of sand, seeing the same lightbulbs, coconuts, and pinecones. And after months of walking the shoreline, DoC ranger Simon Taylor had found only one small piece of ambergris, which he held in the palm of his hand like a strange nut.

"The pigeons at Doughboy Bay and the kakas are as thick as can be," Adamson had told the reporter in 1914. "You don't get much sleep with the teal ducks splashing about all night, and the woodhens holding a corroboree every morning at six." Adamson came from the Shetland Isles, a rugged archipelago that sits in the North Sea to the northwest of Scotland. The wild landscape around Mason Bay would have reminded him of home. Before settling on Stewart Island, he had worked as a tailor's apprentice, a fisherman in Iceland and the Faroe Islands, then came spells in the navy and the merchant service, and a period spent herring fishing, before herding sheep on the Campbell Islands—a remote island chain in the southern ocean, farther to the south, and smaller and even more remote than Stewart Island. To Adamson, a long stint on Stewart Island as a sheep farmer and ambergris collector would have felt like semi-retirement.

"The combers," McIntosh wrote in his 1939 article about Stewart Island ambergris collectors, "are able to class pieces at a glance, and unmatured lumps, whether they be ounces or pounds in weight, are 'graved' in specially prepared plots to age and increase in value. Living close to their fields, these men searched the beaches daily, missing very few specks of ambergris and rejecting such obvious traps for the uninitiated as pieces of putrified fat cast overboard from ships' galleys."

Adamson graved his ambergris. "When Adam left for World War I," Sansom wrote in *The Stewart Islanders*, "he buried all his immature ambergris so that it would mature during his absence and bring a better price. After he returned to Mason Bay—he was decorated while at the war—he stepped off the required number of steps, but was not quite sure of the direction, so the ambergris remains hidden there to this day."

I stand among the marram grass on Doughboy Bay, with a shiny green wall of foliage behind me, and I think of Adam Adamson. In an *Evening Star* article about Doughboy Bay from December 1927, Basil Howard wrote, "The bay was once the haunt of sealers. It has come into prominence recently owing to its connection with Mr. Adam Adamson, 'The Ambergris King,' who recently made it his headquarters. Until the timber for his house arrives, he is living in a cave at the southern entrance to the bay."

The Department of Conservation is using the cave at the moment. Fisherman Phil came before them. Supposedly, in the 1970s, the cave was used by a Japanese tourist who overstayed her visa and lived in it for several weeks. But long before this—before the rangers set up their chairs against the wet back wall, before Fisherman Phil and his wife and their poorly made bed, and before Keiko Agatsuma the overstayer—Adam Adamson had lived in its damp recesses, perhaps even using the cave to store his ambergris, or graving it somewhere in the wet sand of the bay instead, before making his way to Bluff to sell it. And for just a little while, I had stood in the cave too, listening to the water dripping down its walls and taking shelter from the sun there in its dark quiet spaces.

*　　*　　*

As I arrive at the Justcafe Coffee Shop on Main Road one afternoon, proprietor Britt Moore and her friends have just returned from a helicopter flight to Mason Bay to look for ambergris. They found none, but Moore—an American who has lived on Stewart Island for seventeen years—is not disappointed. "We only looked for an hour or so," she says. Moore gives me the names of three Stewart Islanders who are known in Oban for their success in finding ambergris: Cyril Leask, Martin Pepers, and Mark Butler.

"Cyril," says Moore, shaking her head as she steps behind the counter, "he's a funny guy. Supposedly, he found a massive lump of ambergris on Mason Bay, a gross gray slab of it just sitting there. It was worth $100,000 or something. And the funny thing was, people had been walking past it for a week. I think it was in the last ten years or so, but I don't think he'll talk about it. The tax people got on to him about it."

Stewart Island is a small place—cloistered, even. And Oban is smaller still. Everywhere I went, for the five days I was on the island, I asked local residents about ambergris. Arriving early in the morning for my flight to Doughboy Bay, I ask the woman behind the sunlit check-in desk if people ever find ambergris on visits to the west coast. "Occasionally," she says. Trying to sound like I am just making conversation as I wait for my flight, I ask her if she's heard about Cyril Leask finding a piece so large that he sold it and bought a boat with the profits. She pauses. I stare at my reflection in her mirrored aviator sunglasses. Seconds pass. It becomes uncomfortable. Behind me, a child plays with a toy phone on the floor and shouts, "Bwing, bwing, bwing!"

Finally, the woman says, "I think Andrew Leask has found a fair wee bit too," and then turns away, leaving me at the counter with my questions.

An hour later I ask the pilot who flew me to Doughboy Bay about ambergris, shouting over the deafening noise of the propeller as we hurtle through the air and drop over the hills. The day before, while buying groceries in Oban, I'd tried to talk with the proprietor of the well-stocked Ship to Shore General Store, but she soon became overrun with customers. I left, clutching a warm meat pie, and watched the fishing boats returning to the harbor. And a day later, standing in front of a gigantic map of the island, I asked the staff at the ferry terminal on the wharf if they knew anyone on the island who regularly found ambergris. They referred me to Mark Butler. They knew his wife, they said. His full-time occupation was searching for ambergris, they said. In the musty quiet of the Rakiura Museum in Halfmoon Bay a day later, I asked bespectacled research secretary Jo Riksem if she knew anyone on the island who might be willing to speak with me about ambergris. She, like so many other islanders, suggested speaking with Cyril Leask and Mark Butler. "Do they find ambergris often?" I asked. "You'd need to talk to them," she said firmly, and then left me alone with the old newspaper clippings.

Responses were sometimes friendly and more often tight-lipped and suspicious, but no one ever pretended not to know about ambergris. Not on Stewart Island. On Stewart Island, it would have been a conceit to do so—roughly equivalent to Alaskan Inupiats claiming their language has no words to describe snow. This is ambergris country: bleak and windswept and pounded constantly by a rolling, churning surf that carries flotsam from across the Pacific Ocean. I heard the same names again and again: Cyril Leask, Martin Pepers, and Mark Butler. If there is a lesson to learn from my exchanges with locals, it is this: if you spend several months a year walking the remotest and most windswept parts of the west coast of Stewart Island to search for ambergris, your neighbors will probably have heard about it. And, soon enough, I had heard about it too.

When I spoke with Tim TeAika, he had dismissed the efforts of Stewart Island's resident ambergris hunters. "It's just the hopefuls, I think," he had said uncertainly over the phone. "It was a bit of an industry for one or two locals, but I certainly didn't go out of my way to look for it."

If anyone now occupies the wild wet spaces vacated by the men like TeAika, it is these characters: a small group of shrewd, weather-hardened, tight-lipped men willing to endure the conditions of the west coast with a shrug, walking a thousand miles through the rain each winter, cold water sluicing between their shriveled toes for weeks at a time, in search of ambergris. There exists no sense of fraternity even between the ambergris collectors. In fact, they probably distrust and dislike each other more than they do anyone else. They wish one another harm or, at least, bad luck.

TeAika had called them the hopefuls. The Hopefuls. Think about this: it takes several days just to walk from Oban to the west coast. They could fly like I did—take a helicopter or a single-engine Cessna and be there in twenty minutes—but these men wouldn't consider it an option. That would be an extravagance. They could save a day or so—and a lot of money—by taking a water taxi part of the way, and then walking the rest of the way to the coast by a grueling track. Perhaps a water taxi wasn't too much of a concession to the limitations of the human body. And then begin the days, or weeks, or even months, of walking, with their eyes downcast, scanning the beaches, the logjams, the tidal mudflats and river mouths, the bays, the sandbars, and the inlets, in the rain. The Hopefuls? Fisherman Phil lives in a cave for six weeks during the winter, sleeping on a lopsided, homemade bed. Mark Butler searches full-time for ambergris in isolated pockets of the west coast visited by almost no one else. So does Cyril Leask. We sleep, but they are out there searching. They are determined and single-minded to a degree that is almost pathological. If these men are hopeful, this is a new kind of hope, unfamiliar to me.

*   *   *

On my last day on the island, a cold wind begins to blow from the east, swirling around the boats moored in Halfmoon Bay. I walk along the harbor, trying to bury my neck into my collar. Low thin clouds drift over the wharf. The tables and chairs in front of the century-old South Sea Hotel quickly empty, and a bank of solid gray clouds rolls in from the sea, taking up residence in the bay. Suddenly, in the middle of summer, I am reminded of my proximity to the South Pole. I wish I'd brought a coat with me. Easterlies always bring cold weather like this, but westerlies are different. Westerlies bring ambergris. A loose-lipped Stewart Islander would tell you as much, if there were such a thing. Like a 12-mile-long scoop out of the west coast, Mason Bay is beaten every winter by strong westerly storms that dump ambergris on its exposed sands. On Stewart Island, everyone knows this.

A child answers the phone. I ask for Mark Butler and wait. Outside, the sun is shining brightly, but the wind is full of glinting Antarctic menace. The water in the harbor looks like broken glass. I am enjoying the relative comfort of the only public phone box in Oban. It smells of old urine, but it is warm. A large brown moth with the same idea flutters through the air and lands on my shoulder.

A minute or so later, Butler is on the end of the line. He is not eager to talk with me about ambergris. He sounds a little angry. There is something of a precedent: in the recent past, Butler had been involved in the filming of a New Zealand-based television show called *Hunger for the Wild*. In each episode, two chefs drive around New Zealand in a vintage car; at each location, one of them hunts for the main dish and the other gathers the other ingredients. Butler—an accomplished deer hunter— had taken one of the chefs to Mason Bay to hunt whitetail deer. He had received no payment for his involvement and was harboring a deep resentment for his part in the process. He tells me matter-of-factly that he also plans to write his own book on ambergris hunting, which means, he says firmly, it isn't in his interests to talk with reporters like me.

Butler is something of an enigma on the island. He goes by two names: Mark Butler and Mark Moxham. No one can explain why this is the case, but everyone seems willing to indulge him. Several people have spoken about him to me, but none have been kind. People tend to roll their eyes when he is mentioned. He either has several occupations or none at all. When, for the third time, standing in the phone box, he asks me, "What do I get for talking to you?" I answer, approximately, that he might enjoy the feeling of well-being that accompanies accommodating behavior. This, apparently, is not sufficient. I offer to change his name and remove specific information about where or how he finds his ambergris. He refuses. "There are tricks," he says, "and that's what I'd be giving away. It's not always after high tide, and it's not always westerlies. There are a few tricks and, once people get to know them, we're all affected, you know, those of us who make a living off it."

Butler is right, of course, to some degree—assuming there are a lot of people who are willing to fly to Stewart Island, an overgrown, hilly, and half-forgotten dot in a vast and endless ocean, to spend months at a time walking its remote and inhospitable coastline in search of a vanishingly rare fatty concretion of squid beaks produced by just 1 percent of sperm whales.

Several other Stewart Islanders, I tell Butler, have told me that hunting for ambergris is his full-time occupation. "It's pretty much my full-time job during winter," he admits. "I do a lot of walking and stuff. I do about a hundred kilometers a week during winter. Well, over about eight days, I suppose. I mean, Mason Bay is twelve kilometers long."

Outside, it is colder still. The boats are rocking gently on the swell in the harbor. A thin white mist is beginning to form again in Halfmoon Bay, turning the green hills gray. "It's taken me quite a few years to be good at it," Butler continues. "You can go out and find nothing, or just a few small pieces, and then another time you can get lucky. The big pieces are the ones that most people are after. All I can say is, you've got to be there when it's happening."

He pauses. I say nothing. The moth flutters in the phone box. The boats bob in the harbor. "I've probably already said too much, haven't I?" says Butler quietly, almost to himself. And he makes a few excuses, grumbles a little more, and then hangs up.

* * *

The first time I heard the story of Cyril Leask's lucky find, it had sounded like the sort of tall tale that always grows taller in places like Oban. It sounded mythic. It reminded me of Jack and the Beanstalk. The details were half-formed, and they swirled around Halfmoon Bay like a cold easterly, gradually mutating and changing. In other versions, it was Andrew Leask who found the ambergris, and when he sold it, he bought a new engine for his boat. In yet other accounts, it was Cyril Leask again, but the location had changed: it was Mason Bay, or Hellfire Beach farther to the north, or Doughboy Bay to the south. I suspected the story was a product of island community gossip.

With so many variations to the story, I was inclined not to believe any of them at first. But with each retelling, even the most far-fetched stories slowly become believable. These days, Leask hunts for ambergris full-time, I was told, going to great lengths to veil his movements in secrecy in order to stay a step ahead of the other hunters. Supposedly, Adam Adamson built dams, diverting and redirecting rivers to scour the dunes and expose the buried ambergris. And he walked backward along the beach to disguise the direction he was traveling. Perhaps Leask does the same.

In his 1939 article about ambergris for the *New Zealand Railways Magazine*, McIntosh wrote:

> Nowhere does the substance occur more frequently than on Stewart Island, where the beaches of Doughboy Bay, Mason's Bay, Little Hellfire and Big Hellfire are favourite hunting grounds, being exposed to the fury of wind and sea when the roaring southerlies sweep up from the home of the whales—the ice barrier and the Ross Sea. Many notable finds have been made in these localities, chief of which, perhaps, was that of Mr. John Leask, who kicked a boulder in the sand at Mason's Bay and found it to be a lump of ambergris weighing 2,000 oz.

The profits from the 2,000-ounce piece of ambergris that John Leask kicked in Mason Bay—a huge boulder weighing 125 pounds—would have been more than enough to build a new boat.

And in *The Stewart Islanders*, a history of the European settlement of Stewart Island from 1970, Olga Sansom wrote:

> One of the luckiest finds of ambergris on Mason Bay was made at the south end of the beach. Eric Leask and his father walking the beach to Kilbride saw pieces of ambergris of all sizes, fist-sized and smaller, strewn about. They had to get a kerosene tin to collect it more easily. It was obvious that it was a from a sperm whale which had exploded somewhere at sea, as dead whales sometimes do, so Eric's brothers, George and Stanford, decided to take a look at the Hellfire beaches further north. Hellfire had nothing, so they decided to make back to Kilbride. Near Cavalier Creek they noticed a boulder half-buried in the sand. On examination it proved to be a boulder of ambergris, a twenty-five pound nugget and of a goldy-brown colour like tobacco. With the old horse and cart Eric and his father had set out to meet the two boys, so the good find was thrown on board. The nugget was high-grade quality and, along with the smaller pieces of immature stuff, the "catch" for the few days at Mason Bay realized well over 1000 pounds.

Finding large pieces of ambergris seemed to be a Leask family tradition. I decided to talk with Cyril Leask.

"I'd rather not," he said politely, when I consulted the one-page telephone directory and called him one evening. And then, before I could say another word, he was gone.

* * *

Hours earlier, I had left the Department of Conservation rangers to their spraying program on the dunes. I had been excited by seeing Taylor's ambergris and by the prospect of seeing some that belonged to another ranger later. I watched two of them walk south along the bay, their silhouettes growing smaller and smaller until they were swallowed up by the glaring sunshine and the scattered driftwood. With a renewed sense of purpose, I began sifting through the flotsam again, hoping to find a piece of my own. A day before flying to the west coast, I had taken a boat trip with my wife and son, traveling from Halfmoon Bay to Ulva Island, a peaceful nature reserve in Paterson Inlet. On the boat, I had met Ann Pullen, a ranger for the Department of Conservation. White-haired and grandmotherly, Pullen is in her sixties. As we watched sturdy white albatrosses bobbing on the waves near the boat, she had offered me only one piece of advice: "Be careful of the quicksand," she said gravely, "and I'm really serious about that."

I had heard about the quicksand on Doughboy Bay. And it worried me. It had worried my wife, who was sitting in Oban with our one-year-old son. Weeks before I flew to Stewart Island, I had bought several maps. At home, I had unfolded them and spread them over the floor to study their contours and coastlines. One of the maps was a large-scale topographical map of Doughboy Bay, which included, stretching toward the northernmost margins of the map, the southern half of an abbreviated and truncated Mason Bay. Clearly marked on the map, near the spot the pilot had bounced and landed the Cessna at low tide, printed at an angle that followed the gentle curve of the bay, was one word: QUICKSAND.

A river cuts through the middle of Doughboy Bay, rust-colored with tannins from the leaves that choke its bed inland. It was brackish when I first arrived on the beach, but it became fast-moving as the tide began to rise. This, I am guessing, is where the sand is waterlogged and dangerous. This is where the word QUICKSAND is printed on the map. As I approach the channel, my feet sink deeper and deeper into the sand, which makes a sucking sound every time I take a step. The rising water dislodges some of the larger logjams that have formed across its mouth like a poorly made fence. I am standing on a hot cluttered patch of sand, surrounded by a colorful array of tide-swept and incongruous objects that have been broken and worn down by the sea. The tide is too high, the water is too fast, and the sand is too wet. I sink up to my ankles in it. On the other side of the channel, the bay continues, curling around to the south. Arctic terns drift across the sky like slender kites.

All day, I had seen sun-bleached whalebones on the beach—half-buried in sand or tangled in seaweed and plastic bags. Outside the hut, someone had left a long white jawbone with empty tooth sockets leaning against a windowsill, like a stowed umbrella. Next to it, a whale vertebrae, a thick disc of bone with its three processes radiating from it, which reminds me of a naval mine. I find another broken fragment of jawbone, like a pale elbow. Doughboy Bay is a remote and dangerous place, governed by the natural elements.

In October 1998, two hunters camping in the bush arrived on Doughboy Bay and found the beach covered with almost three hundred dead or dying long-finned pilot whales. It was one of the worst whale strandings in recorded history. The large pod had entered the sheltered bay, and the whales had become beached as the tide receded. Early the next morning, DoC rangers helicoptered onto the beach. Hundreds of the whales were already dead, and rangers shot the few that were still alive. The carcasses of 288 long-finned pilot whales were left to decompose on Doughboy Bay.

If I get stuck in quicksand here, no one will help me. I consider the risk of trying to cross the choppy brown water in the channel. Standing in the sun, I have a sudden vision: my green Bushman's hat, on the surface of the quicksand—seaweed stuck to its brim—the only proof that I was ever here. Perhaps the bodies of Adam Adamson and Fisherman Phil, and numerous other long-forgotten ambergris hunters, are still sinking beneath my feet, pirouetting slowly through the soft wet sand, past old graved lumps of ambergris, coming to rest somewhere deep below me in the darkness. I don't want to join them. With a final glance at the southern end of the bay, I turn my back on the channel and its logjams and head north. Retracing my steps in the silence, I kick a few rounded pieces of pumice aside and walk toward the hut. On the way, I lift some of the larger pieces of driftwood, looking for ambergris in their wet shadows.

But I find none.

* * *

Marty "Doc" Pepers has been the district nurse on Stewart Island since 1991. When I first see him in the health clinic, he is wearing blue jeans, a scruffy red T-shirt, and a short pair of gumboots that end just above his ankles. His T-shirt has "Trust Me I'm a Doctor," printed across the chest. He reminds me of a younger Gene Hackman—in his forties, burly and balding, with curly, closely trimmed red hair. At first I assume he is a fisherman. Perhaps he has sustained an injury on one of the fishing vessels moored in the harbor. But then he calls a patient's name, and a boy and his mother rise slowly from their seats and disappear into one of the examination rooms.

"In my life," he says, "I've probably found about maybe half a kilo of ambergris."

I ask him if he has any at the moment. "Bugger all," he says. "I've probably got about thirty or forty grams. I've got a wee cupful, that's it. The biggest piece I've seen was forty-two kilos, about fifteen years ago."

"Was it found on Mason Bay?" I ask him.

"No," he says. "Hellfire, I think it was. This big forty-kilo piece had been pushed out of a whale's arse. You could see it. It was one big huge fucking poo."

"Who found it?"

"Cyril Leask," he says.

His grandfather found a big hunk too, when they were kids. He got two thousand guineas for it and that was a huge amount of money in the thirties. That was like bloody winning the Lotto. He found half of it, and then he went back and found the other half just around the corner. But he's been looking for twenty-five, thirty years. Good on him. And he's a lovely bloke. There are big finds, but they're rare and few and far between. There are a few people make a living out of it, but they're very secretive about it because they don't want everyone going down looking for it. At the end of the day, like, I've walked behind the experts and found it.

None of them will talk to you, you know. Cyril Leask, he won't talk to you. He's too private. A very private man. He's the expert; he's not interested, doesn't want people to know. Mark Butler's one, Mark Moxham, have you spoken with him? He's different. He's a mongrel. It's like the gold rush mentality. You could call it the Ambergris Conspiracy. Well, there is! And the thing is, it's the only part of a whale that you can harvest without causing any harm to the whale.

I ask Pepers if he's heard of Fisherman Phil. "Yeah, yeah," he says.

Him and Butler, they're just new boys on the patch. They can smoke as much as they want when they're away. They're away from their kids. They're on the beaches. They can go hunting. And make money. What a great life. It's pretty good. Money for mud, really.

To me, it's cultural. It's not about money. What I don't like about these new guys, they're too commercialized, looking for it, and there are old fellas here on the island who have been finding it for years as part of their annual income, you know, and they've been doing it since they were kids. Every time I go hunting, I find a bit. I only go over there once or twice a year, but every time I go to Mason's Bay I find some. Yeah, I know what I'm looking for too, you see. The trick is, on a beach, is to look for something that doesn't make sense, okay? Because it can be any bloody color at once, from black to gray to white. It can look like pumice. It can look like snot. I've seen big greasy bits of it that are green. You'd walk past it and think it was rotten seaweed, but it's not: it's ambergris.

So, really, some of it's obvious. It's not a shell, it's not a stick, it's not a piece of rotten seaweed, and that's what you find on the beaches. You find shells, sticks, bits of rotten seaweed, this and that, stones, you know. The thing about Mason's, I've found coconuts on Mason's Bay beach. So where the fuck do they come from?

Why I like having ambergris in the house is every now and again I'll burn a little bit on the fire on a very special occasion, put it on top and get that beautiful smell through the house, you know. And it smells like shit, ambergris, but when you burn it, it smells divine. I'm trying to dry some in the freezer at the moment. I've got a big soft bit. The thing is: How do you age it? How do you age it quickly? That's the conundrum with fresh stuff. How do you get it to go gray because that's where the money is, you see. I've done one trick which worked. I got a piece of fresh ambergris, soft ambergris, and I stuck it in an ice cream container, drilled it full of holes, put sand in it and stuck it on the roof, and just let Mother Nature do its thing. And after six months, half of it had turned white. But I thought okay . . . if you put a piece of blue cod or a piece of fish or a piece of paua in the freezer, what happens to it if it's not well-sealed? It dehydrates. So this is a real soft bit. It's about the size of a human shit, or half the size of a decent human shit, and I've stuck it in the freezer, just for an experiment, to see what it does. It probably won't work but I thought, if there's moisture in there, it's going to start pulling the moisture out of it.

He laughs a dry, humorless laugh, but I know he's not joking.

* * *

On Doughboy Bay, the sun is still hanging high above the water, but the air has grown cooler. The green hills have started to find their shadows again. The plane is an hour late, and Oban seems like a long way away. I have finished my water, and I'm thirsty. I'm disappointed too. I was hoping to find ambergris of my own—to see a large lump of it in the distance, like a misshapen tree stump. But I found none. For an hour or so, I've been waiting in silence in the stuffy one-roomed hut, listening intently for any sound that might slowly gather around itself and grow into the drone of an airplane engine. I stare at the long white whalebones leaning on the windowsill. Several times, I have picked my bag up, opened the door, and then realized I have mistaken the buzzing of a trapped bee for an airplane making its way over the hills.

In the hut, I write and date an entry in the dog-eared visitors' book. Leafing through the previous pages, I see two names I recognize: Robbie and Rob Anderson, the ten-year-old boy who found a piece of ambergris on Long Beach, near Dunedin in May 2006, and his father. In March 2009, almost a year before me, the Andersons had spent several days here, sleeping on the makeshift bunks. Their names are written in blue ink, with the address listed as Longbeach, Otago. I wonder to myself if they found a fist-size piece of ambergris hidden among the wet piles of driftwood on Doughboy Bay.

As I wait for the Cessna to take me back to Oban, I meet DoC rangers Winston Polotu and Jonathan Armitage. Sunburned, they are trudging slowly along the beach after a long day in the dunes. I am reminded of French Foreign Legionnaires. Armitage says he has two more pieces of ambergris. "It's the first time I've ever found any," he says, unrolling a Ziploc bag with two dark shapes inside. "I've been searching for weeks."

"I've got a piece too," says Polotu, opening his large hand and showing me a dark peanut-size piece of ambergris. Polotu is thickset, built like a rugby player, but he looks tired. He is wearing a hat with flaps that protect his ears and neck from the sun. I suggest he'll be happy when this ten-day stint on the beach is over. "Not now we've started finding ambergris," he says.

Back at the DoC hut, Armitage makes himself a drink, shaking rehydration salts into his water bottle. I photograph his two largest pieces of ambergris, placing them on my notebook. The smaller piece is black, like a little lump of coal, and is the length of my thumb. Soft and tacky, it smells like sheep dung. The largest piece is a dirty yellow color. Like Stewart Island, it is "irregularly triquetrous in outline." It is the size of an egg, and its pitted surface is creased and marked with gentle depressions, like a piece of wet clay. It looks like a lump of moldy cheese, mustard-colored and soft. I wonder if this is what cheese might look like if it spent a week or two rolling around in the southern Pacific Ocean. Perhaps someone had thrown it from the wharf in Halfmoon Bay and the currents brought

Ambergris collected by Department of Conservation worker Jonathan Armitage on Doughboy Bay. Credit: Christopher Kemp.

Ambergris consists mainly of cholesterol secreted by the intestinal tract of the sperm whale. This highly waxy substance was often found washed ashore, especially on Mason Bay. Used as a fixative in the perfume industry ambergris was known to fetch $15 an ounce.

whalebone

splicing rope,

ved from
ssel Endeavour
tober 1795.

Ambergris in the Stewart Island Museum collection. Credit: Christopher Kemp.

it here. I lift it to my nose, and the smell is unmistakably complex: like rotten wood and dung and the smell of seaweed, which smells of freshly turned soil and old cabbage. It is ambergris.

In the distance, I hear the thin sound of an airplane rising from the other side of the nearest green ridge.

"There was another guy who used to work for DoC," says Armitage, taking a long swallow from his bottle as I pick up my bag and prepare to leave, "and he sold a jarful of ambergris for $3,000 or $4,000."

8 ON THE ROAD

The reported find of a piece of ambergris on the sea beach at Napier appears to have been somewhat premature. On analysis the ambergris is said to have proved to be tallow. ✴ New Zealand Evening Post (June 13, 1900)

"You would not have known me," remarked the skipper, a polished, courteous gentleman, to the press representative, in his quick, genial way. "Blood from head to foot, but I got the ambergris." ✴ CAPTAIN LARSEN, of the Norwegia, to the Poverty Bay *Herald* (January 1913)

Anton van Helden has a problem. He has lost his ambergris. I watch as he rushes around a spacious well-lit utility room at the Museum of New Zealand Te Papa Tongarewa in Wellington, trying to locate it among the thousands of other specimens and samples stored there. The marine mammals collection manager strides across the room and into another avenue with cardboard box walls. I try to follow him, but at the end of the aisle he makes a tight turn and disappears from sight. When I catch up with him, he is standing with his head down, his chin on his chest, silently thinking. Then he bustles down another aisle, his untucked shirt flapping behind him like a sail, past shelves filled with more mostly unmarked boxes.

It is early in the morning on a wet day in Wellington. Outside, the rain is falling in diagonal sheets on the choppy water. An hour earlier, I had arrived at the harbor-front museum building to discover that van Helden's office is elsewhere, another mile or so across town at a satellite location. I ran uphill in the rain, splashing clumsily through puddles, leaving behind the untidy gray water in the harbor. I am now standing in a multipurpose basement as van Helden continues to search for the ambergris he has agreed to show me. We had entered the room—past mounted deer heads and a stuffed porcupine in a case—and van Helden had nodded fraternally at a man bent in a crouch, carefully pinning and arranging a taxidermied mouse for a museum exhibit.

Blowing stray hairs from his forehead, he fishes a cell phone from his pocket, dials a number, and waits. Stowed incongruously on the shelf behind him, like wood saws in a cluttered workshop, are the long tooth-filled jawbones of marine mammals. No answer. He pockets his cell phone and disappears down another aisle. "No," I hear him say, quieter now in the distance. "No, no, no."

And then: "Ha!" he shouts. "Here it is!"

Moments later he is opening a box and pulling out a folded-over Ziploc bag. At the bottom of the bag is a large black piece of ambergris. Working slowly, van Helden unravels and opens the bag, releasing an expanding nimbus of ambergris fumes into the room. We walk toward a table—van Helden carrying the ambergris. On our way we pass through an invisible but fragrant odor cloud.

He places the ambergris on the tabletop. It resembles a large misshapen bowl of charred wood. In fact, if someone had taken a large wooden salad bowl and then burnt it until it was black and crumbling, it would look a lot like the ambergris in front of us. Leaning over the desk, van Helden uses the tip of a pen to point out several squid beaks, embedded in the rim of the ambergris like shiny black seeds. "All right," he says, "you can see, look, there are just masses of inclusions of squid beaks throughout. It's not like it forms around any particular squid beak. It's not like the reaction that makes a pearl."

We take turns bending toward the desk to smell the ambergris in front of us. When it is my turn, I place my nose in the concavity formed by the curving walls of the blackened bowl and inhale deeply so that I can savor the odor, which seems to have pooled and collected in its uneven base. I ask van Helden to describe its complex aroma, hoping his familiarity with this and other pieces of ambergris will help him to succeed where others have failed. He struggles: "I find it's that sort of rich, kind of tobaccoey kind of smell . . . ," he says, before pausing, searching for words, and beginning again, "A sort of musky, musty . . . kind of tobaccoey . . . almost, yeah, slightly cow patty kind of smell. It's a sort of rich sweet smell." He straightens and thinks for a moment. "There's nothing unpleasant about it," he says finally.

"What we know is that it occurs effectively in the intestine of the sperm whales, where it's produced. So, it probably isn't in the stomach as people often think. It's not thrown up. Quite how it's secreted we don't know."

Measuring perhaps twelve inches across, it is the largest piece of ambergris I have seen. "This was cut out of the gut of a sperm whale," says van Helden, before dipping his head toward the tabletop to smell the ambergris again. "It probably caused the animal to die. It created a peritonitis in the animal. It's about six years old now." I ask him where it was harvested. "I'm not at liberty to tell you that," he says quietly. "It's from the

Bowl-shaped piece of ambergris in the Museum of New Zealand Te Papa Tongarewa. Credit: Christopher Kemp.

New Zealand coast. The animal was a largish female. It was a much, much, much larger chunk: a lump, like, you know, like a big boulder. It totally occluded the gut, so it would have caused terrible problems for the animal."

In fact, the piece that van Helden is now placing back into its Ziploc bag was once the outer layer of the larger piece, like a nutshell wrapped around a nut. The inner core—the nut within the shell—is currently touring the world, van Helden tells me, as part of Whales | Tohorā, an exhibition of whale related specimens jointly curated by Te Papa and the Smithsonian Institute.

I ask van Helden where the rest of the ambergris is. He thinks for a few seconds, with his head tilted to one side. "It's in Pittsburgh," he says. For a moment, we say nothing, standing next to the ambergris, wrapped in the translucent folds of its Ziploc bag. I slowly become aware once again that my feet are wet and my shoes squelch when I walk. After running through the rain across Wellington, I am hoping that van Helden has more pieces of ambergris to show me. "We did have some older pieces," he says, "but quite where Kent put those now is a mystery to me. They should have been all together. But we don't have much."

He looks at me again. "That's the point really," he says.

* * *

A few weeks earlier, we had loaded our beat-up Subaru with all of our be-
longings and hit the road in search of ambergris. Over the next month
or so, we headed steadily north, from Dunedin toward Auckland. Crisscrossing the country, we drove from the east coast to the wild and moun-
tainous west coast and back again—westward across the Southern Alps
and then eastward through narrow passes, dropping into flat basins
braided with the bright milky blue waters of glacial rivers—back and
forth we zigged and zagged, returning to the indented coastline. But always approximately north.

Halfway through our journey, we stood on the sunny windswept deck of
a ferry as it slid across Cook Strait, with our car parked in its hull. Ghostly
constellations of jellyfish drifted past us in the blue water. In September
2008, the cylindrical block of animal fat that washed ashore on Breaker
Bay had passed through these waters first, bobbing toward land on the
unpredictable currents. "It was seen about ten days ago in the middle of
the Cook Strait by one of the ferries," Wellington City Council member
Paul Andrews had told Channel Three News in 2008. "At that stage it was
reported as an iceberg as a bit of a joke, and it subsequently washed up
here." Leaning against the railing of the ferry in the cool sun of early autumn, I scanned the whitecaps for a pale gray boulder of ambergris.

By this time, my son is eighteen months old. He has learned to walk
along the tide line, selecting random objects from the tangle of marine
debris that gathers there. Standing in the surf, he brings each object to
his nose, smells it, and says, "No," in his quiet voice, before dropping it
back onto the sand. In short, we have produced a second-generation ambergris hunter.

* * *

Around this time, I have begun to buy my own ambergris. Weeks earlier, I had purchased a few samples of different grades from Adrienne
and Frans Beuse in Dargaville. Then, trader Peter Chiu in Taipei sent
me two pieces of ambergris after I tell him I plan to make perfume with
it. A third little lump—a single gram—is sent by an anonymous seller
in London, after I buy it on eBay. These pieces come from across the
world in nondescript little envelopes. No one who saw the packets piled
on my desk would have the slightest suspicion that some of them contained samples of decades-old ambergris—whitened lumps of a fatty

concretion of squid beaks that had swirled around vast oceanic gyres for lengths of time that could only be guessed at.

Late one night, I unpack them all, organizing them in a neat and orderly row on my desk. First, I unwrap the pieces that were sent by Peter Chiu. On the envelope is a customs sticker covered with Chinese characters. Inside, there is a fresh black pea-size piece of ambergris, wrapped in a bundle of tissue paper. It has become flattened in transit and is now stuck to the paper—a soft, almost-round disc of pungent, indole-rich, scatolic ambergris. Next to it, I place several fragments of a harder gray sample, which has broken into three or four flinty pieces. Once I remove these flattened, jagged shards, I notice a fine gray dust, which has collected in the creases of the paper like ash.

Alongside these samples, I place a Ziploc bag measuring two inches square. Inside is a dark gray piece of ambergris that came from London. It is approximately the size and shape of a sugar cube. Without removing it from its bag, I can smell it from several feet away. Parts of the inside of the bag are dark and tacky, covered with a sticky ambergris residue. Removing it from the bag, I turn it in my hand like a die to inspect each face of the cube. One face is divided into numerous strata. I trace them with my fingertip: a thin tea-colored seam, sandwiched between two solid black layers.

From two large padded envelopes, I remove several pieces of ambergris sent by Adrienne Beuse. They are each packaged in a little black velvet purse with a gold drawstring. First, a large gray-and-white speckled lump, almost exactly the same size as the top joint of my thumb. On its underside, is a smooth dark concavity that fits perfectly against the pad of my thumb. At one time, this piece was wrapped—like the piece Anton van Helden showed me in Wellington—around a spherical interior, and I am looking now at the point of contact between the two strata. With this piece, Beuse has sent a smaller fragment of white ambergris. I unwrap it and place it carefully next to the others. Shaped like a white domed mushroom cap with a shiny, patinated surface, it's worth more than $100. But it looks like a dirty piece of chalk.

*   *   *

Without doubt, I knew where each of these pieces of ambergris had come from. They had formed—layer upon stratified layer—in the hindgut of a sperm whale and were then expelled at sea, and borne on ocean currents for months and sometimes years.

Back in Wellington, Anton van Helden had been matter-of-fact about where his blackened piece of ambergris had come from. There was no mystery to it. But the true source of ambergris had remained a mystery—at least to most of the world—until as late as 1783. For at least a thousand years, and perhaps much longer, people had been collecting ambergris from the shoreline with no idea of where it had come from. It happened slowly, but gradually the answers came. In 1729 Caspar Neumann, writing in Berlin, had painstakingly collected and listed around twenty of what he considered the likeliest sources of ambergris. They had ranged from the barely believable (the fruit of a tree) to the completely ridiculous (a grounded meteor). Almost as an afterthought, Neumann had included the possibility that ambergris came from whales—a suggestion that, within the context, seemed no more or less ridiculous than any of the other numerous prevailing theories. It gained few supporters and attracted plenty of detractors.

As late as 1773, an edition of the *Encyclopaedia Britannica* still included, in its entry for ambergris, the following: "There has been many different hypotheses concerning the origin of ambergrease, but the most probable is that which supposes it to be a fossil, bitumen, or naphtha, exuding out of the bowels of the earth, in a fluid form, and distilling into the sea, where it hardens, and floats on the surface."

A fossil, bitumen, or naphtha. It was none of them. Slowly, inexorably, the facts began to accumulate, as if they were somehow attracted to one another.

*   *   *

The year was 1721. It was the Age of Enlightenment. A revolution in scientific philosophy was taking place across Europe and in the American colonies, which was based on careful observation, on measurement and empiricism. The city of Boston was in the tightening grip of a smallpox epidemic. Bostonians had everything to gain from a new approach to scientific inquiry.

The outbreak had begun in the spring. The first deaths were reported in May, as the buds began to open on the trees. By October, the disease was claiming more than a hundred lives each week. As the weeks passed, the epidemic continued to spread, jumping from one neighborhood to the next, sweeping in a wide arc across the city. In desperation, Cotton Mather, a prominent local Puritan minister, appealed to Zabdiel Boylston, a local physician. Boylston was already well-known for a series of earlier medical breakthroughs: in 1710 he had become the first American physician to perform a surgical procedure, successfully removing a patient's gallbladder stones; and eight years later, he had performed the first surgical resection of a breast tumor. Mather had achieved a degree of notoriety too, for his involvement twenty-nine years earlier in the Salem Witch Trials. In 1689 Mather had published a monograph titled *Memorable Providences, Relating to Witchcrafts and Possessions*, in which he described the witchcraft of a Boston woman. After a short trial, the woman, who was accused of bewitching the children of a local family, was hanged for her crimes.

Following the publication of Mather's account, paranoia spread across New England, culminating in the Salem Witch Trials in 1692. Hundreds of men and women were accused of witchcraft, and dozens were put on trial. Insinuating himself into court proceedings, Mather encouraged the five judges—three of whom were acquaintances of his—to consider admissible what he called "spectral evidence" in court proceedings. In all, twenty-nine people were convicted of witchcraft, nineteen of whom were later hanged for their crimes. Almost immediately, the governor of Massachusetts, and even the most vocal accusers, realized the executions were an awful miscarriage of justice, and the trials were declared unlawful in 1702.

Somehow, Cotton Mather survived the Salem Witch Trials with his ministerial reputation intact. In the years since, he had learned from his Sudanese slave, Onesimus, a technique that was used successfully to inoculate African villagers against smallpox. As the outbreak of 1721 worsened, he began trying to convince local doctors that the same methods could be used to vaccinate Bostonians. Numerous doctors refused, but Zabdiel Boylston agreed. The technique was primitive. "His custom," wrote William White in 1885, in an anti-vaccination screed titled *The Story of a Great Delusion in a Series of Matter-of-Fact Chapters*, "was to make a couple of incisions in the arms, into which bits of lint dipped in pox-matter were inserted. At the end of twenty-four hours the lint was withdrawn, and the wounds dressed with warm cabbage leaves." Boylston began boldly, inoculating his own sons. Then two slaves. By the middle of December 1721, he had inserted lint coated with pox-matter into the arms of more than 250 Bostonians.

Boylston's career included a second act too, although it was not quite as redemptive as Mather's. In 1724 Boylston made one more astonishing observation. Published in *Philosophical Transactions*, it had the following title: "Ambergris Found in Whales, Communicated by Dr. Boylston of Boston in New-England." It was a bombshell of sorts. And it solved a mystery that had persisted for almost a thousand years: Where does ambergris come from?

Included in the same volume of *Philosophical Transactions* were papers with such enigmatic titles as "Part of a Letter from the Reverend Mr. Wasse, Rector of Aynho in Northhamptonshire, to Dr. Mead, Concerning the Difference in the Height of a Human Body, between Morning and Night," and "Some Observations Made in an Ostrich, Dissected by Order of Sir Hans Sloane, Bart. By Mr. John Ranby, Surgeon. F.R.S." In one superlatively odd paper from November 1725—titled "An Account of a Fork Put up the Anus, That Was Afterwards Drawn out Through the Buttock; Communicated in a Letter to the Publisher, by Mr. Robert Payne, Surgeon at Lowestofft"—the author described the case of a teenage apprentice to a ship carpenter in nearby Great Yarmouth, who attempted a groundbreaking cure for his constipation: "Being costive," wrote Payne, "he put the said Fork up his Fundament, thinking by that Means to help himself, but unfortunately it slipt up so far, that he could not recover it again."

A month or so after the shipbuilder's apprentice put the fork up his fundament, he began to experience pain in his left buttock. For a while, he voided "Purulent matter" on a daily basis. On his buttock, a swelling began to develop. A few days later, when it finally burst, the prongs of a fork were visible through the skin. Cutting around the tines, Payne surgically removed the fork, noting, "It is 6 Inches and a half long, a long Pocket-Fork; the Handle is Ivory, but is dyed of a very dark-brown Colour; the Iron Part is very black and smooth, but not rusty."

*   *   *

Sandwiched between such oddities, Boylston's paper must have seemed almost pedestrian and easily overlooked. It was brief too, taking up less than a single page. Two hundred and six words. But the answer to a millennium-old mystery was contained within it. It began: "The most learned Part of Mankind, are still at a Loss about many Things, even in medical Use; and, particularly, were so in what is called *Ambergris*, until our *Whale* Fishermen of *Nantucket*, in *New-England*, some three or four Years past, made the discovery."

With astonishing brevity, Boylston dismantled the outlandish myths surrounding ambergris, recounting the moment that whalers in Nantucket found it inside a whale carcass: "Cutting up a *Spermaceti* Bull

Whale, they found accidentally in him, about twenty Pound Weight, more or less of that Drug. After which, they, and other such Fishermen, became very curious in searching all such Whales they kill'd; and it has been since found in less Quantities, in several Male Whales of that Kind, and in no other, and that scarcely in one of an Hundred of them." A careful empiricist, Boylston was unwilling to overstate his findings. Instead, he concluded: "*Whether or not (from the Account above) the* Ambergris *be naturally, or accidentally produced in that Fish, I leave to the Learned to determine.*"

It was the answer people had been waiting centuries to hear. Although it was the first description, it was by no means the last. Not long afterward, on April 2, 1725, Paul Dudley wrote a private letter from Roxbury, Massachusetts, to the Royal Society: "I beg the favour that nothing may be published as from the Royal Society about Ambergreese," he wrote, "till you receive my account of that secret, for I have taken no small pains to be a master of that matter, and the Society may depend on the account I shall send them, having got it from the most capable persons in this Country."

Three days later, he sent his disquisition on ambergris to the Royal Society, with the lengthy title: "An Essay upon the Natural History of Whales, with a Particular Account of the Ambergris Found in the Sperma Ceti Whale. In a Letter to the Publisher, from the Honourable Paul Dudley, Esq.; F. R. S." But he was too late. He had been beaten to the punch by Boylston. An associate justice to the superior court of Massachusetts, Dudley was an autodidact. As a member of the Royal Society, he had already published papers in *Philosophical Transactions* on such varied subjects as rattlesnakes, Niagara Falls, earthquakes in New England, and a method to discover "Where the Bees Hive in the Woods, in Order to Get Their Honey."

In every way, Dudley must have felt scooped by Boylston's single-page communication. During a period when the whalers of Nantucket were the world's premier sperm whale hunters, Dudley had spent considerable time with them to research his paper. The publication of both papers, just months apart, was no coincidence. And neither was the involvement of Nantucketers. By 1725, when Boylston and Dudley were publishing their monographs in *Philosophical Transactions*, the whaling industry was growing rapidly in North America. Nantucket had become the most productive whale fishery in the United States and the world. By this time, Nantucketers had begun to venture farther afield in search of whales, leaving behind their shore whaling days, building whaleships to hunt the deeper open waters of the Atlantic Ocean. There, for the first time, they began to encounter sperm whales. In *Leviathan: The History of Whaling in America* (2007), historian Eric Jay Dolin wrote: "For nearly forty years, from roughly 1712 until 1750, Nantucket dominated the offshore whale fishery, primarily targeting sperm whales, but also rights and humpbacks. The number of whaleships on the island rose from six in 1715 to sixty in 1748, and the amount of oil processed annually grew from 600 barrels to 11,250, a nearly twentyfold increase in production."

Naturally enough, in the places where sperm whales were killed and brought to shore, the men who killed them were the first to learn the truth about ambergris. It was a revelation. Fifty years earlier, the natural philosopher Robert Boyle—regarded by many as the first modern chemist—had published his own account in the pages of *Philosophical Transactions*, in which he shared his own theory of the origins of ambergris: "Amber-greece is not the Scum or Excrement of the Whale, &c. but issues from the Root of a Tree, which Tree how far soever it stands on the Land, alwaies shoots forth its roots towards the Sea, seeking the warmth of it, thereby to deliver the fattest Gum that comes out of it."

The intrepid whalers of Nantucket could not have proven Boyle more wrong.

* * *

The weeks tumble by. On a hot day near the end of summer, I drive east for two hours under blue skies, through hills covered with yellow sun-baked grass, from the thermal resort town of Hanmer Springs toward Kaikoura on the eastern coast of the South Island. In the afternoon, I plan to board a boat to take a whale-watching tour.

Kaikoura is a geophysical oddity. It sits on a narrow coastal plain backed by the Seaward Kaikoura Range, a long blade-like wall of snow-capped peaks that runs almost parallel to the coast, clearing heights of eight thousand feet in places. Beneath the forbidding peaks, Kaikoura is a quiet and ordinary place, with around two thousand residents. But less than a mile offshore, the seafloor drops sharply away, forming part of an extraordinary system of deep oceanic trenches and canyons that stretches eastward toward the Chatham Islands, and north all the way to Tonga. Like a deep curving basin, the Kaikoura Canyon sits to the south and southeast of the Kaikoura Peninsula like a fattened horseshoe. In places, it is more than a mile deep. This underwater canyon then extends north, joining a larger system called the Hikurangi Trough, which then runs northward to meet the Kermadec Trench, a 750-mile-long abyssal trench. At depths of more than six miles in places, it is one of the deepest oceanic trenches in the world. Together this sprawling underwater range forms a vast system of bluffs and rises, deep rocky valleys, escarpments, terraces, and mesas that covers thousands of square miles of ocean floor.

Underwater geological features like these are unusual so close to the coastline. In Kaikoura, their effect is specific: just a few miles from shore, they replicate deep-sea conditions. As a result, sperm whales—normally a roving deep-water species—are present here year-round. Tourists visit from across the world to take whale-watching tours. With a backdrop of jagged peaks, they board tour boats operated by a handful of companies, watching as whales spout noisily, breathing in lungsful of air at the surface to prepare for dives into Kaikoura Canyon that might last an hour. Excited at the prospect of seeing sperm whales in their own environment, I park my car in a large lot herringboned with tour buses and make my way through a crowd of tourists to a busy reception desk.

"All the whale-watching tours are canceled today," says an old woman from behind a desk. "They're running seismic exercises today." She looks over my shoulder, to the wide bay. "There won't be any whale activity out there, so we've canceled all the tours."

I ask the woman for more information. Who is running the seismic exercises? She shrugs. It seems inconvenient in an oddly personalized way. I walk slowly back to my car in the sun, trying to decide whether to begin driving the two hours back to Hanmer Springs, where my family is waiting. I drive a mile or so south, to Fyffe House instead.

*   *   *

At the southern end of Kaikoura, Fyffe House sits on a green rise of land, overlooking a long sloping plateau of rock that disappears into the Pacific Ocean beyond it. Painted powder pink, half-hidden by an untended screen of shrubbery, and bordered by a white picket fence, it is all that remains of the Waiopuka whaling fishery, established here in 1842 by European settler Robert Fyfe. The oldest parts of the building—which was built for the cooper who made the barrels for Fyfe's whale oil—were built on pilings of whale vertebrae. I can see them now—pitted white discs of sun-bleached bone, the diameter of dinner plates, almost hidden beneath the pink-painted woodwork.

I walk across the road in front of the house and step onto a sloping rocky table, which extends some distance into the bay, absentmindedly kicking stray stones into the water. The sea is calm, and the sun is so bright that I struggle to look at the gentle swell on the ocean surface for more than a few seconds at a time. Standing before a flat blue bowl of water, I try to discern the vast deep-walled underwater canyon system that lies beneath the surface, stretching for hundreds of miles in all directions. A lone gull flies overhead, flapping white against the sky. I stand on my rocky platform, trying to remain as still as possible. Watching the surface of the ocean for clues, glancing at the gulls overhead for a telltale sign—a strange mid-flap stutter perhaps; or an unusual feint toward the shoreline—I wonder what evidence a seismic exercise would leave for someone standing on the shoreline to witness.

Fyfe and his men were shore whalers. On spotting a whale spouting in the bay, a lookout stationed on the shore would raise the alarm, sending the whalers into a flurry of activity. Using the sandy beach farther along the curve of the bay, they launched their boats in pursuit of the whale. Killed whales were towed back to the rocky platform I am now standing on—which forms a long natural wharf—where they were stripped of their blubber and tried out, a process that turned the water and the rocks bright red with blood. During the 1844 whaling season, Fyfe operated four boats and a crew of thirty-five men from the shore, harvesting 72 tons of oil and 3 tons of whalebone, or baleen. The following year, he took 110 tons of oil and 4.5 tons of whalebone. The year after that, 119 tons of oil, with three boats and fewer men. Arranged on the shelves and mantelpieces inside Fyffe House, like strange dusty bric-a-brac, are fan-like pieces of whalebone and old jars of spermaceti oil. The walls are covered with faded whaling-era photographs. I walk past lace-curtained windows and along quiet unpeopled hallways, papered with peeling Victorian wallpaper.

Back in town, later in the afternoon, I find a large misshapen piece of ambergris in Kaikoura Museum, a modest single-floor building filled with a clutter of artifacts. Stored in a wooden case behind glass, it is long and flattened, like a deflated rugby ball, mostly white and dimpled in places with darker gray patches. Its label reads: "Found north of the Kowhai River[,] Kaikoura. Donor: Bev Elliott."

It is the second piece of ambergris I have seen since we left Dunedin. We are still a week away from van Helden in Wellington. Days earlier, we had seen another sample at the Okains Bay Maori and Colonial Museum, a sprawling collection of reconstructed historic buildings almost hidden in a green coastal fold of the Banks Peninsula, two hours east of Christchurch.

It was a rainy day. I walked from one ramshackle outhouse to another, past cabinets filled with an array of stone adzes, bone fishhooks, and

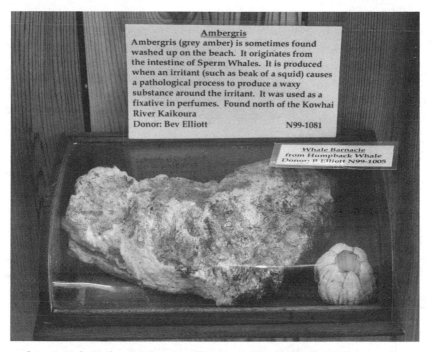

Ambergris in the Kaikoura Museum collection. Credit: Christopher Kemp.

samples of indigenous basket-weaving. The museum and the surrounding buildings were deserted. In a small hut near the entrance, I was met with a wall of glass eyes: stuffed ducks, weasels, and deer; fur seal and dolphin skulls—all arranged in a crowded cabinet.

Outside, rain poured noisily from blocked gutters, forming puddles in the grassy meadow. A windmill turned in a gray blur. An unlatched door banged against its peeling frame. The stuffed ducks glowered in the settling silence. The place was deserted. I photographed a long, slender wooden Maori war canoe from 1867 and a collection of stone adzes and primitive tools. Everything seemed to be decaying. Baskets slowly unraveled in the damp coastal air. It was an eerie place. A large spacious room, the Whare Taonga was filled with thousands of Maori artifacts. A man's voice on a muffled vocal track spoke to no one as the wind whistled and the rain splattered outside.

In the farthest corner of a large hall cluttered with exhibits, next to a bright green to-scale model of the pleated Banks Peninsula coastline, I found a display cabinet dedicated to the New Zealand whaling era. There, next to scrimshawed sperm whale teeth and a rusting spade-ended blubber knife, was a large black brick of ambergris. A hand-cut label was tucked beneath it, as if it was a strange and fragrant paperweight. The label, written in black pen, reads simply: AMBERGRIS.

* * *

Finally, on February 13, 1783, any remaining confusion surrounding the origin of ambergris was laid to rest in London. Joseph Banks presented a paper to the gathered members of the Royal Society, written by Franz Xavier Schwediawer, a German physician living in London, and titled "An Account of Ambergrise":

> We are told by all writers on ambergrise that sometimes claws and beaks of birds, feathers of birds, parts of vegetables, shells, fish, and bones of fish, are found in the middle of it, or variously mixed with it; but of a large quantity of pieces which I have seen, and which I have carefully examined, I have found none that contained any such thing, though I do not deny, that such substances may sometimes be found in it; but the circumstance which to me seems to be the most remarkable, is, that in all the pieces of ambergrise of any considerable size, whether found on the sea, or in the whale, which I have seen, I have constantly found a considerable quantity of black spots, which, after the most careful examination, appear to be the beaks of the *Sepia Octopedia*. These beaks seem to be the substances which have hitherto been always mistaken for claws or beaks of birds, or for shells.

Schwediawer had managed to put together a collection of several ambergris samples, all of which contained well-preserved squid beaks, and displayed them for members of the Royal Society. Alongside them: a cuttlefish beak that Schwediawer had removed from a preserved specimen in Banks's collection.

"Any gentleman, who will be at the pains to compare them together," wrote Schwediawer, "will be enabled to convince himself of the truth of what I have advanced."

And so the truth was incontrovertibly known. For a select few, this mattered, but for most, the revelation meant little.

Even more impressive though was Schwediawer's intuitive understanding of what his findings meant. More than 220 years before Robert Clarke made the same assertion, Schwediawer wrote:

> The beak of the Sepia is a black horny substance, and therefore passes undigested through the stomach and into the intestinal canal, where it is mixed with the faeces; after which it is either evacuated with them, or if these latter

be preternaturally retained, forms concretions with them, which render the animal sick and torpid, and produce an obstipation, which ends either in an abscess of the abdomen, as has been frequently observed, or becomes fatal to the animal; whence in both the cases, on the bursting of its belly, that hardened substance, known under the name of ambergrise, is found swimming on the sea, or thrown upon the coast.

*   *   *

It is midmorning and Wilma Blom meets me beneath the airy dome of the Auckland War Memorial Museum's spacious Grand Atrium. A small crowd has begun to gather, milling around the cafeteria entrance, waiting for the museum to open. Blom and I leave them behind, ascending the stairs that spiral gently upward, following the sweeping exterior walls of the atrium. I have come here to see a piece of ambergris. In fact, I have come here to see AK75784: a large, rounded piece of ambergris that weighs slightly more than two pounds and has been part of the museum collection for almost twenty years.

Once upstairs, we enter a wet and dry storage section of the marine collection, walking past shelves of blond wood filled with jars of starfish and mollusk species. I am reminded vaguely of an upscale hair salon. Blom introduces me to Tom Trnski, the research manager, gray-haired and fit-looking, dressed in jeans and a green T-shirt. Together now, we all exit the storage room and continue down a curving corridor toward Blom's office. The moment we enter her office, I smell it. AK75784. It rests on Blom's desk, mottled with black and white patches, tethered by a thin length of black twine to a piece of wood that has been painted pink.

Found in February 1992 on Ruapuke Beach, a black-sanded beach a hundred miles south of Auckland, situated along the sinuous welt of the west coast, the ambergris was donated to the museum by the married couple—"a Mr. and Mrs. Suisted," Blom tells me—who found it there.

A week earlier I had pulled on a wetsuit, hopping gracelessly along the dark silty sand at Ngarunui Beach, a few miles north of Ruapuke Beach, so that I could float in the frigid ocean on the west coast. For an hour, I drifted between the surfers. In almost complete silence, I floated on my back watching the grassy cliffs as they yawed up and down, framed on either side by my feet. For long seconds at a time, I was invisible to everyone on the beach, lying at the bottom of blue valleys of shining water. Minutes earlier, preparing to enter the surf, I had accidentally picked up my wife's wetsuit instead of mine. My wife is much smaller than I am.

For several minutes, I had wrestled with the suit—tugging at the foamy material and grunting with effort—only realizing my error when it stuck firmly on my hips and would not move higher or lower. The black neoprene was stretched across my thighs, taut and gray.

Exasperated, I had pulled the suit off, picked up the larger one and mistakenly—hastily, and to my horror—put my feet into its sleeves. I stumbled around on the sand, like a strange insect, with the black legs of the suit twirling in the air. By the time I finally lay in the water, I had almost exactly no dignity left. And I was exhausted. Wisps of cirrus clouds drifted high above me in the blue sky. I lay in the ocean and gave myself to the water. I slipped over the crests of the breaking waves, imagining that I was an ocean-borne block of ambergris. Sturdy black-backed gulls flew across the sky. The only sound I could hear was the static-like rush of sand particles colliding with each other in the water, propelled along the ocean bed by the waves.

Back on Blom's desk: a dark, almost spherical, core sits in the middle of the lump. At one end, additional gray strata wrap like a rind around the curving exterior of the core—an exterior layer almost swallowing a spherical interior core. Parts of it are covered with a thin white cracked crust, like ash. From certain angles, the external strata are shaped like a snake head with the dark inner core resembling an egg, being swallowed. In places, its shape becomes more irregular and knobbly, with lots of smooth, rounded protrusions. It is approximately the size of a misshaped bowling ball. By now, the odor is unmistakable: a sweet pleasant earthy rush of tobacco and old wood that lingers in the nostrils, accompanied by a faint tang of the ocean and the slippery mudflats at low tide.

I ask Blom where the ambergris is stored when she isn't showing it to people like me. "Oh, right there," she says, nodding over my shoulder to a conspicuously empty space on an otherwise well-utilized shelf. "I don't know whether we've all got used to it," she says, "but I really don't notice the smell anymore."

9 GONE A-WHALING

*Occasionally visitors to Cape Cod have heard ambergris mentioned by griz-
zled whalers of New Bedford. But the public in general never heard of it.* ∗
The Practical Druggist (1922)

*They call me the "Ambergris King," and I'll tell you why they call me that.
You see, I know all about it. Like anything else, a man's got to be trained to
ambergris.* ∗ DAVID C. STULL, talking to author William Tripp in
There Goes Flukes (1938)

"We're supposed to have a twenty-pound block," says Michael Dyer
loudly from the top of a stepladder, "and I just have no idea what happened to it."

Dyer, a curator of maritime history at the New Bedford Whaling Museum in Massachusetts, is teetering on the top step of the ladder, surveying the least-visited shelves of a tall storage unit. He is looking for ambergris. I am standing beneath him, trying to extrapolate—given Dyer's
posture and everything I know about gravity—exactly where he will land
when he falls.

"It might actually still be in the museum somewhere," he continues,
now searching another shelf. "It's not in here, but then we have at least
one other storage facility that I have not been through entirely."

It is early December and I have come to New Bedford, driving an hour
south of Boston to the south Massachusetts coast, hoping to see more ambergris. There is a reason I am here and not elsewhere. For more than a
hundred years—from the mid-eighteenth century until the beginning of
the twentieth century—New Bedford was the world center of the sperm
whaling industry. From the New Bedford Harbor, vessels embarked on
cruises that took them around the world to hunt sperm whales—to the
Arctic and the Antarctic, to the Pacific, the Azores, and the Indian Ocean.

At its peak in 1857, the New Bedford whaling fleet consisted of 329
vessels employing more than ten thousand men, making it one of the
wealthiest cities in the world. In a black-and-white photograph from
1870, the Central Wharf of New Bedford is completely covered with barrels of whale oil. The geometric repetition of the barrels, arranged in
neat rows along the harbor front, seems at first to be an optical illusion.
One has to stop for a moment and look at it again to understand that the

wharf is hidden beneath a thousand wooden barrels laid end to end, like oversize cobblestones. Used to make candles and as a fuel for lamps, the oil became such an important commodity that New Bedford was known as the "City That Lit the World." And so if ambergris has an ancestral American home, it is New Bedford, Massachusetts. For more than a century, much of the ambergris that was found anywhere in the world was brought here first.

Dyer picks up a rough-hewn wooden box, filled with several dark brown rectangular bricks. "I think this is whale oil soap," he says, holding one of the heavy blocks in his hand. He brings it to his nose and inhales. "It's soap," he says. "Believe it or not, that stuff is really good soap. It really does the job. It lathers up beautifully and leaves your skin nice and soft and cleans exceptionally well, and I know that because there were some little flakes of it and in the process of putting stuff on exhibit I grabbed one of those flakes."

He holds his thumb and index finger an inch apart. "It was just about that big," he says, "and I took it down and I washed my hands with it. It was made in New Bedford in 1835 by Zenas Whittimore. It's this big block of this hideous-looking stuff. It looked like that. But, boy, was it nice. Nice soap. Smelled like soap, worked just like soap. It actually says it right on there: Whittimore. It's stamped on the soap itself." He places the box of whale oil soap to one side, looks up with a grimace at the shelving in front of us, and scratches his salt-and-pepper hair.

*   *   *

In the spacious lobby of the museum, three huge whale skeletons hang in the air, suspended more than forty feet above the ground. Stretching almost the length of the building are the enormous pale bones of a juvenile blue whale (*Balaenoptera musculus*) skeleton. More than sixty-five feet long—only about half the length of an adult specimen—everything about it is still oversize. The slender, slightly flattened ribs descend from the vertebral column like warped pieces of driftwood, forming the curved walls of a rib cage that is large enough to sit inside. The elegantly bowed jawbones of the lower jaw protrude into the empty air like an implausibly long shovel. Above it, the elongated skull is pointed and birdlike. It measures almost twenty feet in length and weighs one and a half tons.

The oil-stained skull continues slowly to release oil, filling the air of

the lobby with its fumes. It oozes from the surfaces of the large porous
bones. Dripping at intervals onto a well-placed plastic tray, the oil enters
a long snakelike coil of plastic tubing and descends to ground level, collecting there in a glass flask. Half an hour earlier, I had placed my nose at
the mouth of the flask, sampling the odor of the whale oil. More than an
inch deep and dark brown like maple syrup, it is intensely aromatic and
reminds me of the thick slippery smell of diesel. Cetologists expect the
bones to continue releasing oil for at least the next sixty years.

On an upper level of the museum, the center of another room is filled
by a sperm whale skeleton. I walk around it, admiring its enormous pale
flabellate scapulae, which measure several feet across. Its long, thin bottom jaw is studded with a row of incurving conical teeth. The skull—
whalers called it the sleigh—is shaped like a narrow trough. It reminds
me of an old and weather-beaten wooden boat. At its center is a smooth,
rounded concavity—the cistern that once housed the mysterious spermaceti organ.

*   *   *

It is cold in the storage facility. Dyer is on his knees now with his head
buried in a large box of miscellaneous whaling artifacts and specimens.
He picks up a little glass jar, quickly inspects it, and places it back into the
box. "That's all Japanese," he says to himself quietly, moving to another
box. He picks up another specimen. "That's all Japanese," he continues.
"Japanese. Japanese."

I look at bottles of honey-colored whale oil and the cracked green
paint of ships' boxes, arranged in an orderly row on the shelf nearest to
me. "We're interested in it, you know," Dyer had said when I first asked
him about ambergris, "but we're interested in collecting. If there's nothing to collect, then it's simply information. Information-wise, you've got
chapters of stories on ambergris. Who found it and how they found it. To
start with, *Moby-Dick*, obviously . . . that chunk of the book is an important part of the story. But there are many other accounts of substantial
chunks of this stuff that came back to New Bedford, and they were duly
entered as material obtained in the fishery."

We are trying to locate several pieces of ambergris that belong to the
museum collection. "We have a number from the Atlantic, for sure," says
Dyer, "late in the 1880s, 1890s, in that time frame. For some reason or
other, a lot of ambergris came back and was recorded at that time."

Ambergris in the New Bedford Whaling Museum collection, in New Bedford, Massachusetts. Credit: Christopher Kemp.

He pushes the stepladder toward another shelf and begins to climb it. I watch as his head disappears between boxes and trays. "Okay," he says then, in a quiet and muffled voice. "There we go."

Moments later Dyer steps off the ladder with a large rusted tin in his hands. On the outside of the pockmarked tin is a label and printed on it are the words: "M. L. Barrett & Co. - Importers & Manufacturers - Fine Drugs, Essential Oils, Vanilla Beans, etc. - Chicago." Dyer opens the tin and pulls out a large black piece of ambergris, the size of an apple. "Now that's ambergris," he says, holding it toward me. "Smell it. It smells like ambergris. What I don't see prominently are any squid beaks in it. I wonder whether it has been processed somehow."

He places the ambergris on a desk and pulls out another smaller piece with a shiny whorled surface. "It looks like charcoal and smells like tobacco," says Dyer. "That's the way I describe it." In fact, it has a much more complex odor than that. It smell it and I'm reminded of tobacco, old musk and furniture polish, and the sort of thick cloying fragrances that grandmothers tend to wear. As I inhale, I become aware of an entire spectrum of different odors, some of which I can identify, while others pass over my olfactory bulb and diminish too quickly for me to recognize. Lurking somewhere deep beneath the immediately accessible surface aroma, I can detect the incongruous smell of black licorice, which lingers stubbornly in my nose and stays on my hands for hours. "I'd put that between 1900 and 1920, the first twenty years of the twentieth century," says Dyer.

During the long years spent in storage, perhaps a hundred years or more, a black powder has accumulated at the bottom of the tin. It is an inch deep and thick like coal dust.

"That's some pretty nice-looking stuff, right there," says Dyer, pouring the powder out onto a carefully placed piece of cardboard. "So this is the real deal. This might answer your question. What did they do with it? This stuff was obviously sold to M. L. Barrett and Company in Chicago, right? It's not a New Bedford thing. It's not a Nantucket thing. It's not a New England thing. So, there's a market for the stuff. It's pretty old. I mean, it could have been as early as the 1880s. It wouldn't surprise me in the least."

I find a little hooked fragment of squid beak and put it to one side. And then another. Dyer finds one too. We begin to line them up carefully in a row. They look like sharp black pieces of broken nutshell. It is an odd moment. For a while, we stand side by side in the silence of the storage room, wordlessly removing squid beaks from a heap of black powder with the tips of our fingers and placing them with the others. In another box, Dyer finds a small glass vial and hands it to me. It is empty. Its interior walls are dusty and yellowed with time. "The pieces that were in here are on exhibit," he says, "but that was collected on board the schooner *John R. Manta* in the nineteen teens, nineteen twenties."

Several times a year, says Dyer, people contact him in the hope that he will confirm that the substance they have sent him is genuine ambergris. He receives these packages of unknown substances—fragments, nuggets, shavings, slices, splinters, and shards—and is unsure of what to do with them. "Every time somebody finds something on the beach," he says, "they think it's ambergris and they bring it in. You know: Is this ambergris? It has to be ambergris. I've never had a piece where I've said, 'Yep, that's exactly like the stuff that I've seen, and that is in fact ambergris.' I've never been able to actually say that. They bring all kinds of stuff. I've got two samples up in my office, and I'm going to start keeping them more and more because it's really interesting. I'll show you the stuff up in my office."

*   *   *

An hour later, and a block or so across New Bedford, Dyer is sitting in the museum's research library in front of a tall gray filing cabinet. Each drawer is filled with index cards, reducing the two thousand or so whaling logbooks in the museum collection into a list of alphabetized terms that can be searched by researchers.

Dyer opens the drawer marked A–B and riffles the top of the index cards with his thumb. Each card represents a reference in a logbook. "You can see," he says, "compared to cards for things like 'blackfish' and 'bone,' just how few there are for 'ambergris.'" In fact, there are only perhaps twenty or so index cards for ambergris, and as many as several hundred cards for more common words like "baleen." An entry from the 1869 logbook of the *Sea Fox*, a whaling bark from Westport, Massachusetts, reads: "Ambergris sold at Nossi bay, Madagascar, for $5,839.00 French gold." From the logbook of the *Marcella*, which detailed a voyage that began in 1873 and lasted three years: "Ambergris sold at Zanzibar for $67. per lb." In another logbook—from an 1879 cruise taken by the *Falcon* of New Bedford—simply: "Ambergris, 136 lbs."

A few weeks before my visit, I had bought *The History of the American Whale Fishery* by Alexander Starbuck, an exhaustive 800-page compendium of whaling records that was first published in 1878. From it, one can learn the ephemera of countless whaling voyages that sailed from American ports between 1784 and 1876: almost a hundred years of whaling cruises, painstakingly tabulated along with dates of departure and return, the name of the captain of each ship, the port of origin, and the amounts of whale oil and bone harvested.

On June 6, 1817, the *Atlas*, a ship from Nantucket captained by William Easton, returned from the Pacific Ocean with 1,372 barrels of sperm whale oil in its hold; in 1841, Captain Prince Sherman of the *Parker* from New Bedford was killed when his whaleboat was stove in by a whale; and in 1876, the crew of the *Camilla* from New Bedford abandoned her in the Arctic, leaving on board 190 barrels of spermaceti, 300 barrels of whale oil, and 5,000 pounds of whalebone. There are thousands of entries like these, for hundreds of ships. The record seems complete.

But I was surprised to find that ambergris is mentioned just five times in *The History of the American Whale Fishery*. And, even in those few scattered instances, the entries are unremarkable: on August 28, 1867, the *Wm Wilson* returned to port with eight pounds of ambergris aboard; in 1842, the *America*, from Wareham, Massachusetts, had brought home eighteen pounds. Modest amounts of ambergris, collected after years spent at sea. And these dates are important: in 1852, in almost the middle of this time period, the New England whaling fleet enjoyed its most successful year ever, killing more than 8,000 whales, from which it harvested 103,000 barrels of sperm oil and 260,000 barrels of whale oil. But still, ambergris is, for the most part, oddly absent from the record.

"The thousands of logbooks that I've read over the years," says Dyer, "and I don't find references to ambergris very often. You see it sometimes. Occasionally. But I mean these guys are killing anywhere from twenty-five to sixty whales between a twelve-month and a four-year voyage, and you don't see it very often. Say you had officers or sailors who joined the vessel somewhere out there in the world and rose through the ranks, they might not actually even know what it is, or have ever heard of it. But I suspect that career whalemen knew all about it."

I sit quietly in the wooden stillness of the library for an hour, lost in the smell of foxed books. A row of old wooden paddles from whaleboats and a long iron harpoon lean against the wall near the entrance, like a collection of arcane relics. Dyer had piled a stack of battered textbooks and whaling logbooks on the table in front of me: *A Compleat History of Druggs* by Pierre Pomet from 1712; a slim and dusty first printing of Caspar Neumann's 1729 dissertation on ambergris, printed in the original German; a handful of whaling logbooks that mention ambergris; and *There Goes Flukes* by William Henry Tripp, published in 1938.

I am sitting a couple of blocks from the harbor. Once such a bustling place, it is quieter now. I had visited it earlier in the morning, before the museum opened. Now a fleet of fishing boats—hulls covered with blisters of rust—rolls on the oily water. In the library, I am hunched over the crinkly blue pages of the *Sea Fox* logbook from 1869, trying to make sense of the faded, looping script. I turn a page and slowly read the following entry, from 1870:

Saturday 10th

At 6 PM got underway and stood to the N.E. ward toward the island Nos-Beh where we came to anchor at noon on Sunday where we lay till the 20th. Obtained some refreshment and stores such Sugar Beans and Rice. Sold all the dry goods belonging to the ship, also those belonging to my self, all at renumerative prices. Also the ambergris, suppose I did realize for it as much as the owners of it expected but I am satisfied with the sale and am glad it is off my hands. It was a bad lot, scarsely a pound of good ambergris in the whole and it had decreased fifteen pounds in weight since it was put on board. After spending a week in negotiations, I sold the lot for the sum of five thousand eight hundred and thirtynine dollars French gold & silver. Waited four days for the money, finaly got it and have it on board together with the proceeds of sales of goods amounting in all to a little over six thousand dollars.

The weather is very hot and if we get clear without sicknes I shall be glad. I shall remit the money to the owners as fast as I can find an opportunity not infringing upon my own voyage. Better to sell ambergris when it is first taken.

*   *   *

Two days after visiting New Bedford, I am standing at the tip of Cape Cod, a long, spiraling strip of land that extends into the windswept Atlantic Ocean. It is just before sunrise on Herring Cove Beach, and the wind is brutally cold. A fierce northerly sweeps in relentlessly from the sea, as sharp as a knife. I lean into it and begin to walk northward, dark gray land rising to my right and silver-edged ocean to my left. It's almost seven o'clock. The sun finally peeps over the darkened Massachusetts landmass to the east, across the water of Cape Cod Bay. It hits the top of the dunes. The sky is a clear wintry pink. In the distance, I can see Race Point lighthouse. The sea crashes up the sloping shoreline, sending a white wall of spray into the air. It is so cold, and my eyes are watering so much that I close them, opting instead to trudge blindly over the tangles of bladder wrack with my nose dripping into the wind.

I am completely alone on a brittle Atlantic winter coastline. One of the most popular beaches on Cape Cod in the summertime, Herring Cove Beach is a challenging and unwelcoming place in December. There are parking spaces for several hundred cars. But, just after dawn, mine is the only car parked there. I never saw more than ten other people the entire time I was in Provincetown. The buildings along Commercial Street, the imposing Town Hall, the stores, and the little houses in the East End were all unoccupied. Plenty of cars lined the streets, but there was no one around to drive them. After an hour or so, the absence of other people became unsettling.

If Cape Cod resembles a muscleman's flexed arm, jutting out from the east coast before bending northward into the Atlantic Ocean, Provincetown sits nestled within the semi-protective curve of its curled fist. The cape stands like a lone bracket. The day before, I had driven south and east from Boston, through quiet wintering New England fishing towns like Barnstable and Yarmouth, before turning north at Harwich to follow the green curving land of the peninsula.

In the seventeenth century, early settlers sent back to England astonishing stories of seas thick with whales. In fact, the abundance of whales was apparent the moment the pilgrims steered the *Mayflower* toward the tip of Cape Cod in November 1620. Writing in *Leviathan: The History of Whaling in America*, historian Eric Jay Dolin described the pilgrims' arrival: "On the morning of November 11, the weary wanderers dropped anchor in modern-day Provincetown Harbor. Almost immediately, whales surrounded the ship."

Before long, the pilgrims began to understand that the whales represented an important potential source of revenue, something they desperately needed. Within a hundred years or so, whalers were beginning to develop techniques that would bring them thousands of whales from across the world. But before the whaling era, the waters around New England were filled with whales. "An entry from a diary written in 1762," wrote Dolin, "tells the story of a sixty-year-old Truro resident who recalled when, as a younger man, 'he had seen as many whales in Cape Cod [Provincetown] harbor at one time as would have made a bridge from the end of the Cape to Truro shore, which is seven miles across and could require two thousand whales.'"

I had arrived in Provincetown after nightfall. A storm was approaching, closing in quickly on the cape in the darkness. The wind rumbled and howled through the streets, rattling flagpoles and shaking Christmas decorations. Lights swung in the wet sky. Hastily handwritten signs on store windows read: "*See You in the Spring!*" In the darkness, I walked through Whaler's Wharf, an enclosed parade of shuttered candy shops and souvenir stores, and stepped onto the sandy beach of Provincetown Harbor. Twenty feet away, huge breakers crashed on the beach. Two small boats were moored at the shoreline. Each enormous wave threatened to break them into pieces. First their prows pointed to the night sky, and then they were gone altogether, hidden behind another towering wave

Gulls whipped over my head, ghostly in the night. I crouched low near the sand and made my way along the curve of the bay in the darkness, trying to detect the smell of ambergris on the shore before the winds stripped it away from me.

But it was a hopeless endeavor.

*   *   *

Settlers to New England were not the first to benefit from colonizing a new and untapped source of ambergris. For centuries, ambergris had been an important footnote in numerous accounts of conquest and colonization. In October 1613, when the fledgling colony of Bermuda was still named for its founder, Admiral Sir George Somers, John Chamberlain wrote in a letter to Sir Dudley Carleton:

> Great store of ambergris from the Somers Islands this year, the only commodity as yet. People begin to nestle and plant there very handsomely. The Spaniards, nothing pleased thereat, threaten to remove them next year, but the inhabitants are nothing dismayed, trusting rather to the difficulty of access, than to any other strength of their own. A piece of ambergris found as big as the body of a giant, the head and one arm wanting, but so foolishly handled, that it brake in pieces. The largest piece brought home, was not above 68 ounces, which sells for 12 or 15 shillings an ounce more than smaller pieces.

In 1682, in *Carolina; or, A Description of the Present State of that Country,* Thomas Ashe wrote, "Ambergrise is often thrown on their Shoars; a pretious Commodity to him who finds it, if Native and pure, in Worth and Value It surpasses Gold; being estimated at 5 and 6 Pound the Ounce, if not adulterated." In 1678, when the Lords Proprietors of Carolina appointed Robert Holden as the collector of customs for the new colony of Carolina, history had already taught them an important lesson: ambergris was a valuable substance. It deserved well-worded legislation. With its long and unbroken Atlantic coastline, the new colony of Carolina produced a lot of ambergris.

In a letter dated February 19, 1679, the Lords Proprietors issued Holden the following instructions: "You are responsible for wrecks, ambergris, and other 'ejections of the sea,' as well as rents, and will receive 10 per cent. of all receipts and recoveries for your pay." Similarly, in 1703, when James Moore became "the Receiver Generall of all that Part of our Province of Carolina that lyes south and west of Cape Feare," he also received instructions to "take into your Possession our share of wrecks, Ambergreen and all such other things as of right belong to us."

*   *   *

For several decades at the beginning of the twentieth century, the entire North American ambergris market—in fact, the world market—was

ruled by one man: David C. Stull, the self-proclaimed "Ambergris King."
He lived in Provincetown, in a handsome little house on Commercial
Street that looked out on the harbor. Before his death in 1926, he claimed
he had bought and traded at least half of the ambergris that had ever entered the United States.

Stull's watch oil business, which he ran from a cramped refinery in
the East End of Provincetown, was his first concern: "Mr. Stull conducts a
trade that is unique," reported the *Washington Post* in April 1911. "He pur-
chases an animal oil that is worth $15 a gallon aboard the whaleship and
by a secret process converts it into an oil which sells readily at the rate of
25 cents for a thimbleful."

Although his watch oil business clearly paid the rent, Stull's true obsession was ambergris. As early as the 1880s, he had bought a ten-pound
lump of ambergris, obtained in the south Cuban keys by the crew of a
Provincetown schooner called the *William A. Crozier*, and then sold it for
$500 per pound. It was an astonishing sum. For the next four decades,
Stull handled more ambergris than anyone in the world.

"It is said," reported the *Post*, "that less than one and a half tons of
ambergris is known to have been offered for sale in the history of the
world, and that Mr. Stull, in his capacity of agent for a wealthy French
concern, has handled half of that quantity. Moreover, he has obtained
nearly all of the precious stuff brought to port by the dwindling fleets
of sperm-whale seekers in the last twenty years, thus possessing a giant's
share of the world's supply."

As his reputation grew, Stull—whose business card read: "First Hands
for Ambergris"—was asked to appraise a constant stream of pieces of suspected ambergris at his Provincetown headquarters. People arrived from
across New England carrying boulders of animal fat, lumps of wood pulp,
wax, and whale blubber.

In *There Goes Flukes*, published in 1938, William H. Tripp included the
following description of a meeting with Stull in 1921. The two met in New
Bedford. Stull had arrived from Provincetown to bid on a haul of ambergris weighing more than seventy pounds, which had been brought to port
by the *Valkyria*:

> "See what I have here," he said as he called my attention to a ring on one
> of the fingers of his left hand. Set in the ring was a large rectangular stone,
> known as a goldstone. Much to my surprise, with a finger nail he raised the
> goldstone, as one would the cover of a box, and inside the setting of the ring
> was a depression filled with a black, sticky substance.

"That's ambergris," he said, "but of very poor quality. Some of a small lot
I bought once. It was no good because it would never dry as it should. I keep
that in my ring to show people what poor ambergris looks like."

I have a photograph of Stull on my desk as I write this. On a sepia-
tinted postcard from the New Bedford Whaling Museum collection, Stull
cuts a strange figure, crouched on the sand in Provincetown Harbor, a
stout little man with white muttonchops that hang down almost to his
collar. He is cutting a deep circular incision around the muscular neck
of a dead and stranded pilot whale. Behind him, a wooden pier marches
into the harbor. A caption on the card reads: "D. C. Stull. Provincetown,
Mass., cutting up Blackfish to manufacture his Watch and Clock Oil."

"A few years ago," reported the Boston Daily Globe in 1909, "there was
great excitement on the North shore over a greasy mass of stuff which
floated onto the beach and was promptly spotted by the wise ones as ambergris. As usual a sample was sent to Mr. Stull, who with cold conclusive-
ness returned it as refuse from the Boston dump."

* * *

Back on Herring Cove Beach, the sand has been scoured clean by the
trailing edges of the storm front that barreled across the cape the previous night. In this bright place, it should be impossible not to stand for
a moment in the wintry morning light to consider the sublime beauty
of water in motion. But it is so cold that I can only withstand it for a
few minutes. I hide in the shelter of the locked restrooms, fumbling to
put on a pair of gloves, and then start to walk back along the sand to
my car. Driving back into town, I return to Whaler's Wharf and walk
the thin strip of sand again. It is cold and sunny now. I find two dead
horseshoe crabs on the sand half-hidden in a long wet frilly drift of
sea lettuce. The storm has thrown up mussel shells, gloves, pieces of
plastic, strips of bubble wrap, fish bones, and a condom. The two boats
I had watched the night before, being thrown around by the waves, are
submerged now in the harbor, broken in pieces, their hulls filled with
swirling water.

Leaving the harbor behind, I walk east along Commercial Street and
find a small nondescript wooden building, its white paint peeling. Now
an art gallery, it was once the site of Stull's whale oil refinery. At one time,
most of the ambergris that entered the United States was brought here,
appraised and graded by Stull, and prepared for the French perfumery

market. No trace of that trade exists. The gallery is closed and filled with bad art.

A little farther along the road, facing the blue glinting water in the harbor, is the handsome white clapboard house that Stull lived in until his death. I linger outside and take photographs of a bright blue plaque on the wall: *David C. Stull 1844–1926 Whale Oil Refiner "Ambergris King" Lived Here.*

Later that day, I follow the arc of Cape Cod back to Boston, stopping along the way to search other beaches after it has become warm enough to walk the shoreline. At Head of the Meadow Beach on the Atlantic coast near Truro, I struggle over grassy and gorse-covered dunes to walk several miles along the coast, my shoes slipping on the steeply inclined sand. Two seals—black-headed and glistening—bob around in the surf, playing among the blue peaks of a twelve-foot swell. At First Encounter Beach near Eastham, I look out across an undulating dun-colored mile of mudflats at low tide. In December 1620, the Pilgrims skirmished with Wampanoag Indians on the sand here.

Dolin described the encounter in *Leviathan*: "A dozen or so Indians were 'busie about a blacke thing,' which the explorers could not quite discern. Seeing the white men approach, the Indians ran back and forth as if gathering their things, and then disappeared into the woods."

The "blacke thing" was a pilot whale, a species that frequently strands on shorelines in huge numbers. In some instances, pods of several hundred pilot whales have stranded at once and beaten each other to death on the shore with their tails. These events confound cetologists, who have struggled for centuries to explain their cause. In this particular instance, the carcass was worth a second glance. Both the Indians and the settlers knew that a single pilot whale could produce about a barrel of valuable oil. In the morning, the Pilgrims and the Wampanoag Indians skirmished on the shore here, exchanging arrows and musket fire.

It is strange to think of the fighting that had taken place here over a dead pilot whale carcass almost four hundred years before. I stand on the parking lot and look out over the mudflats in the winter sun. A few scattered stands of pine are green in the sunlight. Walking onto the slick mud—which is still wet from the last tide—I sink into the thick ooze and almost fall over. I walk back to my car, pausing to wipe my shoes on the yellow dune grass. The sky is blue.

In the parking lot, a lone seagull pecks disconsolately at a dead eel, tugging at the skin around its gray wrist-thick neck.

* * *

"The museum was founded by Louis Agassiz in 1859," Judy Chupasko
tells me as we walk through the Mammalogy Department of the Harvard University Museum of Comparative Zoology in Cambridge, Massachusetts—a sprawling, cabinet-filled Victorian-era suite of rooms that
houses a collection of around eighty-seven thousand mammalian specimens. "It's the second oldest natural history museum in this country actually. It's older than the Smithsonian and the American Museum, and
the Field Museum in Chicago. In the Mammalogy Department, we have
specimens that were collected by Agassiz that date back to the 1830s, actually, so even before the museum opened. He was collecting for a long
time before he got the money to build this place."

We walk into a room filled with row after row of military-gray lockers,
each filled with drawers of specimens. "This room is all rodents," says cu-
ratorial assistant Chupasko, opening one of the shallow drawers. Inside
are two neat rows of misshapen carcasses: North American flying squirrels. I count almost thirty of them, splay-legged and frozen in mid-flight.
"We have about thirty-five thousand rodents in the collections. We have
a good bat collection too. Rodents and bats are the biggest orders of mammals, under the class *Mammalia.*"

I look at the squirrels in the drawer again: a strange sight in their two
rows, white-eyed and lumpy. In front of each specimen is a sturdy little
plastic box, which houses its bones. They had each, in effect, been carefully dismantled.

"They're not taxidermied," Chupasko tells me, when she sees me looking at the squirrels arranged in their rows. "These are called scientific
study skins. People get confused between specimens mounted for display
and education and exhibit, versus scientific specimens. We don't need
them to look pretty, although we want them to look as nice as possible.
The main thing is to make sure they're not greasy, or have meat, or they're
not rotten. That way, we don't attract pests in the collection."

A few feet farther along, Chupasko opens another drawer and pulls
out the stiff, flattened carcass of an enormous Chinese flying squirrel.
With its long bushy tail, it measures perhaps four feet or more in length
and is a deep chestnut-red color. Although they are the size of house cats,
these specimens are a species of flying squirrel too. "These are squirrels found in China," Chupasko explains, holding one in the air. "They're
beautiful too. For a mammal, this is a brilliant color, because most mam-

mals are brown. Most mammals are nocturnal, so they're not as colorful as birds are."

Together, we wander through the department, stopping to look at lion pelts, preserved anteaters and pangolins, and mounted warthog heads. Everywhere we walk, beady black eyes follow our progress. "It's all skeletons in here," Chupasko says as we enter another room. "All the hoofed mammals, mostly the artiodactyls that are in the family Bovidae, like all the cows and antelopes and sheep and goats and bighorn. Those are all in the family Bovidae. But we have the Giraffidae here too and the Antilocaprodae, which are the American pronghorns, so we have three families of hoofed mammals in this room."

After a while, just thinking about the number of different mammal specimens that surround us becomes a dizzying task. With every step, I walk past the jumbled bones of entire orders of mammals. For a biologist, it is like visiting a library that contains a copy of every book ever written. Chupasko points out another two rows of cases. Inside them is the rabbit collection. "The Lagomorpha, that whole order," she says, before pointing in another direction. "We have the pig family and the deer family over here, and the rhinos and hippos over here."

I peer through the distortions and imperfections of an original Victorian hand-blown pane of glass and see inside a jumble of graying bones. Beneath them, a faded card reads: "*Hippopotamus amphibius.*" In another drawer: "*Choeropsis liberiensis,*" or the bones of a pygmy hippopotamus. We walk along another unremarkable-looking corridor, entering another spacious room. Family after family of carnivores unspool beside us with every step. "We call it the carnivore room," says Chupasko, "because, by sheer volume, it's mostly carnivores. It has the cat family, dog family, seal families in it, all the carnivores, bears, mongooses, civets, genets. All the carnivores are in here, but we also have the marsupials in here, and our monotremes, the little duck-billed platypuses and the spiny echidnas are in this room. So we have these, the marsupials, the anteaters, aardvarks, sloths, armadillos—we have a lot of orders in this room but we call it the carnivore room."

Finally, she stops in front of an oversize whale skull, stored upright. It is the size of a small car. The underside of the enormous jawbone is stained black—sooty like charred wood—by old whale oil. "An odd fin whale," she says. "This should be at the field station, but it's just hard to get out of the department. I think the building was built around it or something."

Ambergris, collected by J. Henry Blake and donated to the Harvard University Museum of Comparative Zoology. Credit: Christopher Kemp.

Later Judy Chupasko holds a small green cardboard box in her hands. She places it on a table outside her office.

"You hit a little gold mine here, I think," she says cheerfully, gesturing toward the box on the table before leaving me alone with its contents. I sit and pry open the lid. Inside, there is a haphazard clutter of glass vials and specimen jars from the first decades of the twentieth century. A few have lost their cork stoppers, which have dried and crumbled in the eighty years since they last were stoppered. All of the pieces of ambergris inside the vials came from one man—an artist from Somerville, Massachusetts, named J. Henry Blake.

There is a whitened flat piece from May 1921 that sits on a wad of cotton wool in a fragile thin-walled vial. Next to it, a heavier, ridged glass flask contains a dark lump, like a nut. Its label reads: "Immature ambergris of little value from Mr. Stull." Another vial contains a handwritten label in shiny pencil that reads: "Ambergris best quality 300 pr lb (the piece this came from was 75 lbs and sold 90 pr lb)." I retrieve another glass tube, filled with a handful of black curving squid beaks. Beside it I place a squat glass flask filled with several gray aromatic lumps. I smell the lumps and wince before reading the faded label: "D. Stull 'Ambergris King' says Sewer Grease, From Beach Nausett, Mass. Aug 3, 1915. J. Henry Blake." Working slowly, I remove one container after another.

Before too long, I have arranged in front of me a miniature cityscape made of glass.

*   *   *

On June, 24, 1937, from his home in Somerville, Massachusetts, just a couple of miles north of the Harvard campus, J. Henry Blake wrote the following letter to G. M. Allen, the curator of the mammal collection at the Museum of Comparative Zoology:

> Dear Dr. Allen,
>
> According to your letter of the 17th, I am sending box of specimens of Ambergris and (separate cover) notes, clippings, photographs and list.
>
> It is difficult to get a collection of this nature, because the parties are more interested in the money part than science. Fortunately, I have known Mr. Stull, the "Ambergris King," since boyhood and this has aided me in getting spms.
>
> I am glad this has come into such good hands as your own, and glad to help in any way.
>
> With thanks to Dr. Barbour and yourself, I am,
>
> Very truly yours,
> J. Henry Blake

Inside the brown envelope addressed to Allen—it has become as soft as a bedsheet after decades in storage—there are several beautifully rendered hand-drawn sketches by Blake. Each was drawn on a postcard-size card: a piece of ambergris with a squid beak embedded in it; a square-shaped gray-green lump of ambergris, its surface covered with hairs; a squid eye, removed from a piece of Stull's ambergris. I place them carefully side by side on the table. From the envelope, I remove a black-and-white photograph of a large dark boulder of ambergris. On the back is written: "Taken by Brig *Viola* in 1917, on coast Africa from a 15 bbl sperm whale. I saw at office Howe, French Co. 99 Broad St. Bos. It weighed 162 lbs in Bos. but 155 lbs in NY at which weight it sold for $225. pr. lb. = $34,875."

Alongside the sketches and the specimens and Blake's spidery handwritten notes, there is a yellowing stack of newspaper clippings. One of them, dated November 8, 1911, and taken from the *Gloucester Times*, recounted the washing ashore of a large dead sperm whale at Ocean City, on the New Jersey shoreline.

"The whale was a great show for visitors for a time and then Nature

> J. HENRY BLAKE
> 61 COLLEGE AVE.
> WEST SOMERVILLE, MASS.
> June 24, 1937
>
> Dear Dr. Allen:
>   According to your letter of the 17th inst. I am sending box of specimens of Ambergris and (separate cover) notes, clippings, photographs and list.
>   It is difficult to get a collection of this nature, because the parties are more interested in the money part than science. Fortunately I have known Mr. Stull, the "ambergris King", since boyhood and this has aided me in getting spmo.
>   I am glad this has come into such good hands as your own, and glad "help in any way.
>   With thanks to Dr. Barbour and yourself, I am,
>   Very truly yours,
>   J. Henry Blake

Letter from J. Henry Blake to G. M. Allen (1937). Credit: Christopher Kemp.

began to assert itself," the *Times* reported, "and then the town offered $300 to anybody who would remove the dead carcass. Mr. Whale was 56 feet long and weighed about 30 tons. It had Ocean City by the nose." According to the clipped article, on hearing of the offer, an old New Bedford whaler boarded a train to Ocean City and, in a move that baffled some of those present, paid $500 to secure the $300 contract. "Then, with warp, tackle, cutting irons, and boiling pots," the article continued, "the New

Bedford man went to work. Strips of the heavy blubber came off briskly and went into the pots. Gradually the atmosphere of Ocean City cleared in one way, but darkened in another, for the natives saw that Yankee visitor chop out ambergris by the chunk and boil out oil by the barrel."

Working against time, and in front of an increasingly hostile audience, the whaler finally removed ambergris worth $4,800 from the carcass, along with $1,000 in sperm oil and $500 of crude oil from its tried-out blubber. I sit at the table next to Chupasko's office, reading this and other clippings, and leafing through Blake's copious notes. It is raining outside. I am in no hurry to leave. Through the windows, I can see students crossing the wet streets under umbrellas. Next door, Chupasko types on her keyboard and listens to the radio. I hold each chipped glass vial to my nose for a second and then a third time, savoring the individual odor profile of each fragrant hundred-year-old piece of ambergris. I rattle the squid beaks in their little jar.

It seems to me an error—an error of judgment and omission—that these pieces of ambergris have not been cataloged and described as carefully as the lumpy squadrons of flying squirrels in the nearby specimen rooms, which have been dismantled and categorized, and laid out side by side in their shallow drawer. Their bones have been measured, their stomach contents assessed, the number of embryos in both the left and right horns of the uterus have been counted. They have been reduced to statistical data.

The few pieces of ambergris, crumbling in their jars on the table outside Chupasko's office, represent a glaring gap in our attempts to describe the natural world. The record is incomplete, inadequate with respect to ambergris. From the largest whale to the smallest shrew, everything else in the collection—a staggering total of eighty-seven thousand specimens—has been painstakingly studied, described, weighed, numbered, inventoried, cataloged, added to databases, referenced, and then cross-referenced. Whereas the samples of ambergris—one of the only physical manifestations of one of the most reclusive animals on the planet—have been relegated to a distant storage facility and left to gather dust, as their corks slowly desiccate into powder. The same is true elsewhere: the Smithsonian Institution has one small piece of ambergris in its collection, known simply as USNM 571430; the Field Museum of Natural History in Chicago, with a collection of more than twenty-one million natural history specimens, has none.

There must have been, it seems to me, a moment in the development of scientific thinking when ambergris could have been included in the record. Perhaps it might then have been studied and described as exhaustively as everything else in the museum collection. There must have been an opportunity, but despite Blake's insistence, it has not been taken. Instead, ambergris remains an oddity—a strange relic. One by one, I return each sample to the green box. I stack the paperwork—Blake's clippings and photographs, handwritten notes and sketches—in a yellowing pile and slowly ease it back into its envelope, so thin now that it begins to come apart between my fingers when I reopen it. Arranging it all in the center of the table, like a cluttered shrine to missed opportunity, I thank Chupasko and step reluctantly into the rain.

*   *   *

"Those are *Crepidula fornicata*," says Eric Jay Dolin, the author of *Leviathan: The History of Whaling in America*, pointing to an empty pinkish-brown speckled shell lying open like a book, its inner surface shining whitely on the sand. "The common slipper shell, they're protandric hermaphrodites, so they change sex during their lifetimes. They live one on top of the other in a stack. The ones on the top are the males; the ones in the middle are hermaphrodites; and the ones at the bottom are the females. As you move higher in the stack, you change from one sex to the other."

I am standing with Dolin on Revere Beach, a few miles north of Boston. It is early in the morning and very cold. With his back to the surf, Dolin is holding a large pale shell, which almost fills his gloved hand. It is a large fat shell, with a rounded body whorl. We find another and another—each larger than a golf ball. They are moon snails, says Dolin, who looks enough like Robert De Niro that it seems odd to be standing on the beach with him in the winter, identifying mollusks by their Latin names. He picks up another. "Here's one that recently died," he says, pointing to a stiffened, chocolate-brown piece of tissue on its underside, blocking its aperture. "That's the operculum. That's like the trapdoor, the protective mechanism."

As it happened, I had chosen the right person to invite on an ambergris hunt. We follow the shoreline, and Dolin names every mollusk and each type of seaweed that we find on the sand. "I worked in the Mollusk Department at the Museum of Comparative Zoology at Harvard," he explains. "It was when I was in college. I took some time off. It was in the summer. I had a great time. I worked for Dr. Ruth Turner, who was

With Eric Jay Dolin on Revere Beach in Boston. Credit: Christopher Kemp.

fairly famous—she was one of the first women to be tenured at Harvard, and she went down in the *Alvin* to the deep-sea vents with Bob Ballard. She's the one who did a lot of the scientific work to identify those chemosynthetic tubeworms they found down there. I had a lot of fun. I basically spent the whole summer of 1980 identifying shells from Australia, with my arms in formaldehyde."

The sky is clear and cloudless, lit with a wintry brightness that makes us both squint. We walk between two headlands—Winthrop to the south and Nahant to the north—alongside the flat calm of water of Broad Sound. Farther out to sea, a distant ship travels across Massachusetts Bay. Dolin points to a dark reddish-green bed of seaweed on the shore. "There's *Chondrus crispus*, or Irish moss," he says, "which is a seaweed from which you get carrageenan—a thickener for ice cream. They used to collect it around here for that. See those green fronds? That's *Zostera marina*. That's eelgrass. There are eelgrass beds off here."

I had invited Dolin here hoping that, in some imperceptible way, his presence might somehow be beneficial in my search for ambergris. A couple of years ago, he published a definitive history of whaling in America, which means he is familiar with ambergris. And he is local, living a few miles farther north of here. But, says Dolin, in the eight years that he has lived nearby, he has never walked on the beach before and has only ever driven past it. Established in 1896, Revere Beach was the first public beach in the United States. In 1882 the *Stranger's Illustrated Guide to Boston and Its Suburbs* included the following description of the shoreline: "This magnificent beach is about five miles long, and is lined, at short distances, with hotels, restaurants, cottages and bathhouses. Being but a short distance from Boston, it has always been a favorite resort for the Massachusetts public, and visited during the hot season by the thousands. On a pleasant Sunday, it is not uncommon to see from fifteen thousand to twenty thousand people strolling along the beach."

This morning there are just two of us strolling along the beach. And we are cold. But Dolin is undeterred. He points a leather-gloved finger at three or four large dark, saucer-shaped shells, each as big as one of my shoes. They are, he tells me, surf clams. "*Spisula solidissima*," he continues. "They live in a little bit deeper water. There was a big storm actually, recently. There were some pretty big waves, and that's usually when these get thrown up here. This is *Mytilus edulis*—that's the common blue mussel. When you go to a local restaurant and you order mussels, that's the type that you'll get.

"There just seemed to be this era, a couple of centuries, where everybody was amazed, confused, scared of, and making all sorts of strange conjecture about everything in the natural world," says Dolin, who researched and wrote about ambergris for *Leviathan*. "It was a time of real flux where the understanding was minimal, the guesses were all over the place, and it was fascinating."

Trudging northward through the marine litter, which has been thrown on the shore by the recent storm, Dolin bends and points to an empty pointillated crab shell, half-buried in the sand. "This is the carapace, or the shell, of an *Ovalipes ocellatus*," he says. "It's the lady crab, a beautiful crab around here. They don't have blue crab here, but they have those, and they have Jonah crabs." We crunch over skate egg cases, jackknife clams, mussels, and softshell clams. I pick up a dog whelk—a large shell with a curved rounded body whorl and an elegant pointed spire, which reminds me of a treble clef. "You don't normally see these either," says Dolin. "This one's been beaten up a lot."

We have been walking along the beach for more than half an hour. Finally, we are beginning to feel the effects of the frigid temperatures. My fingers are numb. Turning around at the northern end of the beach, we begin to walk back to where our cars are parked. "Just walking along here," says Dolin, "this brings me back to my childhood because, as a kid, my nickname was Nature Boy. I spent a lot of time alone. My dad was a scientist, a physicist, so I tended to read actually a lot of *Scientific American*. I didn't read many books when I was a kid. We lived on Long Island, near the shores of Long Island Sound, and I loved just walking along the beach, going to tide pools, seeing what I could find."

As my feet crunch over broken shells, Dolin suddenly comes to a stop. "Look at that!" he says, bending toward an olive-green tangle of eelgrass and jackknife clam shells. We are huddled together against the wind on the shoreline. In the distance, cars trundle along Revere Beach Boulevard. I think for a moment that Dolin has spotted some ambergris on the beach. But he points instead to a delicate pearly shell that sits open like a butterfly on the sand. "That's a Purplish Tagelus," he says enthusiastically. "It's one of the more beautiful New England bivalves, with those faint purplish rays of color radiating out toward its edges."

Finally, we have made our way back to where we began. My pockets are filled with moon snail shells and a few long straight jackknife clams. I'm holding the white spindle of a whelk, like a central spoke, which is all that remains after the sea has eroded its Baroque rounded outer spirals. I climb into my rental car, relieved to be out of the cold. Although I have failed, now on the shores of two oceans, to find my own ambergris, I have finally seen plenty of it. There are traces of ambergris everywhere I have been in coastal New England. I leave Dolin standing on the sidewalk on Revere Beach Boulevard, next to the entrance of Kelly's Roast Beef, and drive my rental car southward, away from the shoreline and toward the parking lots and the gray towers of the airport.

10 A MEETING

Adrienne Beuse bends in her chair to pick up a black tote bag from the
floor and places it gently in her lap. She carefully reaches a hand into its
dark folds and pulls out a smaller cloth bag, which she places gently in the
middle of a glass-topped table. In turn, several smaller cotton-wrapped
lumps are removed from the bag. One by one, like a museum curator unpacking artifacts, Beuse unwraps each of them: first, a few small pieces of
rounded white ambergris, like broken little pieces of chalk. "This is like
perfection in white," she says, holding one between her thumb and fore-
finger. "It's about as good as it gets."

Next to it, she places a larger piece, pitted and gray like an egg-size
lump of eroded coral. "That's when it's almost gone past its best," she
says. "It gets pitted like that, and that normally means it's been locked up
in some sand dunes for a while." Beuse then shows me a thin flat piece,
turning it over in her hand like flint. "We call this plate ambergris," she
tells me quietly, placing it with a clicking sound on the table, alongside
the pieces already there. Then she unwraps a few larger, darker pieces,
which join the growing collection. "Classic gray," Beuse says almost to
herself, like a butterfly collector identifying a familiar species.

Slowly, the table becomes cluttered with an array of different pieces of
ambergris. Beuse begins to arrange them in order, from the large granite-
colored pieces to the smaller shiny white fragments, which sit in a cairn,
like shards of bleached bone. A thick sweet aromatic cloud envelops us.

＊   ＊   ＊

Beuse and I are sitting across from one another in a cramped Auckland
hotel room. It is dark. The curtains are closed. Outside, it is a warm Friday

Different grades of ambergris, shown to the author by ambergris dealer Adrienne Beuse. Credit: Christopher Kemp.

A selection of Adrienne Beuse's finest white ambergris. Credit: Christopher Kemp.

afternoon in spring. The curtains are outlined with a border of bright sunlight. Earlier in the morning, from my north-facing window, I had watched a long line of cars moving across the harbor bridge in the distance, toward the green-gray haze of the north shore, windshields winking in the morning light. A flotilla of tiny boats floated in the harbor—a thicket of white masts in the sun.

A couple of days ago, I had taken a two-hour flight north from Dunedin to meet Beuse, who has driven two hours south today from her home in Dargaville. The meeting almost never took place at all. We had tentatively arranged to meet months earlier by e-mail. Ten days before I was due to arrive in Auckland, I sent Beuse another e-mail to confirm our meeting, but she never responded. A few days later, I e-mailed again: still no response. I phoned and left several voice-mail messages, none of which were returned. And then, a couple of days before I left for Auckland, I made a final halfhearted attempt to contact Beuse, and she answered the phone.

"We don't normally like to do the coffee shop thing," she had said, once she'd confirmed the meeting would still take place. "When you pull out samples of ambergris, which is a pretty funny-looking thing, you tend to attract some attention. If you have a hotel room, that's fine."

*   *   *

Middle-aged and a little overweight, Beuse's untended dark brown hair falls to her shoulders. She has thick bangs that curl over her forehead, partly obscuring her eyes. Sometimes, when she blinks, her bangs twitch. "I'd rather you didn't tape me," she says, frowning at the digital recorder in my hand. "I think that will make me guarded in what I say," she explains guardedly. I'm not allowed to photograph her either, she says. I explain that photographs of the ambergris would be useful to me later when I try to describe them in a meaningful way. She agrees: I am permitted to photograph the ambergris on the table, but not Beuse. Every time I raise my camera, she jumps up from her chair and walks stiffly away from the table, like someone walking off a sudden leg cramp. She seems nervous and flustered. During our meeting, she receives text messages, which she answers furtively. And when I handle the ambergris—cradling it in my hands, feeling its weight, slowly raising it to my nostrils—she scrutinizes me. I feel like I'm making a drug deal. In a city the size of Auckland, with more than a million inhabitants, it's probably much easier to buy illegal drugs than ambergris.

"The only time people are really secretive is when there are really big quantities of ambergris," Beuse says. "They tend to disappear into the ether, shall we say, very quickly. If you found twenty kilos and told twenty people, someone would come along and knock you on the head. I mean: those twenty people might each tell twenty people, and then suddenly a lot of people know ... and *then* you might get knocked on the head. People who find a high-value item tend to be very frightened and nervous. If you're buying a piece of ambergris under those circumstances, you're going to those meetings with a lot of cash."

In the months before we finally meet each other, Beuse had offered several times to introduce me to one of the collectors from whom she regularly buys ambergris. I was hoping to accompany one of them on a hike into the isolated and remote country in the north, in search of ambergris. But they had all refused to meet me. Instead, I had this one opportunity. It was an aberration—as if a door had opened for a moment, allowing a glimpse into a secretive and clandestine world. I had managed to slip through the door before it closed again. Beuse's resting state is one of suspicion and guardedness. At the same time, she is here, and somehow not here at all. This feels like only half a meeting: I can speak with Beuse, but I'm not allowed to photograph her or record her voice. For a few moments, I can cradle and inspect the ambergris on the table, but I can never know certain specific and, to me, more important details—like where they were found, or by whom, or when—because she refuses to tell me.

Paradoxically, in other ways, Beuse is one of the most high-profile people in the ambergris world: in 2006, when Dorothy Ferreira received the strange green disc-shaped object in the mail from her sister, the *New York Times* used Beuse as a source for the article. And in 2009, after Ben Marsh swam into a rounded lump of ambergris floating in the ocean near Oakura, Beuse was the trader who first appraised it and then, eventually, purchased it. When I spoke with French trader Bernard Perrin, as he drove through Normandy, the way he talked about Beuse gave the impression that she was an old acquaintance of his. He had, he told me, once bought a few pieces of ambergris from Beuse. And Mark Butler on Stewart Island knows Beuse too, occasionally selling her pieces of ambergris that he finds there.

Together, Butler, Beuse, and Perrin form one small part of a sprawling and tangled network that moves ambergris around the world, from finder to end user. Perhaps Butler, on one of his extended hikes along the remote western beaches of Stewart Island, found an exceptionally high-quality piece of white ambergris and sold it to Beuse in Dargaville, who then sold it to Perrin on the Côte d'Azur.

I lift my camera to take another photo. Beuse rises from her chair again, as if stung by a bee.

*   *   *

On February 6, 1685, Charles II, king of England and Scotland, died at Whitehall Palace, famously apologizing to his courtiers on his death-bed, "I am sorry, gentleman, for being such an unconscionable time a-dying."

Four days earlier, he had suffered what appeared to be a seizure or a stroke. For a short time, he had been unable to speak. Since then, his physicians—as many as twelve of them fretting around him at once—had become more creative. In all, before he died, they subjected him to an estimated fifty-eight remedies. From *Royal Charles* by Antonia Fraser: He was bled several times; doctors gave him white hellebore root to make him sneeze; plasters of Burgundy pitch were applied to his feet; he was given anti-spasmodic medications made from black cherry water and spirits of human skull; doctors administered emetics to make him vomit and gave him enemas of rock salt and syrup of buckthorn; and red-hot irons were applied to his shaved scalp and his feet. Not surprisingly, he died anyway.

Immediately after his death, rumors began to circulate regarding the cause of his seizure. A growing list of theories was collected by Thomas Babington Macaulay in *The History of England*: "His Majesty's tongue had swelled to the size of a neat's tongue. A cake of deleterious powder had been found in his brain. There were blue spots on his breast. There were black spots on his shoulder."

Others believed the Duchess of Portsmouth was responsible. She had given the king, they claimed, a cup of poisoned chocolate. Perhaps, others said, it was the queen who had killed him, with a jar of poisoned dried pears. There was another possibility: "Something," Macaulay wrote, "had been put into his favorite dish of eggs and ambergris."

In fact, Charles II was not alone in enjoying an occasional plate of eggs and ambergris. At this time in history, ambergris was one of the scarcest and costliest substances in the world. It came to England slowly, harvested from remote places, like the Somali coastline and the Nicobar Islands, and then brought westward along the well-trodden spice routes. Over previous centuries, ambergris had come to represent an elevated status especially to the privileged aristocracy, who gladly ate it.

In seventeenth-century England, the Tudors enjoyed an early version of marmalade made of pippin apples and perfumed with ambergris and musk. It was rumored that the Earl of Carlisle served a pie that was flavored with ambergris, musk, and magisterial of pearl. It was so expensive that no one could afford to eat it. "There is no knowing how scientifically a great cook may have distributed his musk and his ambergrease," wrote a contributor to *The Book of Table-Talk* in 1847, "but, not having tasted of such a dish, we are inclined to say, generally, that we should prefer a small

Perigord pie, scented with truffles, and which may be bought in perfection for about two pounds sterling."

Elizabeth I, who ascended to the throne in 1558, was more than partial to ambergris: "To the last Princess of the House of Tudor," reported *The Court Journal: Court Circular and Fashionable Gazette* in 1833, "we are to attribute the universal adoption of ambergris and spikenard, in preference to the more pungent spices in vogue among her predecessors. Elizabeth, an idolator of scents and essences, chose that even the delicate flavour of a quail or ortolan should be lost in a 'steam of thick distilled perfumes'; and but for the frugal predilection of her Caledonian successor for cock-a-leekie, and a singed sheep's head, this feminine corruption of taste might perhaps have been stereotyped into a national infirmity."

Cookbooks from the period are filled with numerous elaborate recipes that require ambergris. The 1685 edition of Robert May's *The Accomplisht Cook* included a recipe for a hen made of puff pastry, with its wings displayed, sitting on top of a clutch of pastry eggs, each of which contained a fat nightingale, seasoned with pepper and ambergris. Alongside recipes for carp pie and fritters of sheep feet, *The Compleat Cook* by Nathan Brook from 1658 had instructions for several dishes that required ambergris, including a partridge tart, and boiled cream with codlings.

To know and experience ambergris as completely as possible, I decide I need to eat it. I begin to search for recipes that I can replicate at home.

*   *   *

Adrienne and Frans Beuse have been trading ambergris since 2004. The previous four or five years, Frans had been an independent ambergris collector. Before that, they owned and managed a hostel together.

"We have probably fifty-plus collectors on our books," Adrienne says. "Most traders will have dedicated collectors who, for one reason or another, will stick like glue to the trader they work with. But some collectors are freewheelers. They'll be with one person one year and then another person the next. They do flit about."

Beuse classifies the ambergris collectors on her payroll: "There are four different types of collector," she tells me.

There's the person who just happened to take a walk on the beach and found some. They just noticed something odd about it and picked it up. They might take it home and put it away and keep it for a while. The next would be someone who knows about ambergris a bit. They go out to the beach occasionally, they know what they're looking for, and they're really what we call sort of Sunday afternoon drivers. And then there are the full-time, part-time hobby people, and they usually live by the sea. They could be farming on the side, or hubby is working but wife goes for a look around in the week. They're hobbyists. Some of them can be very serious: it pays for the bills or the holiday each year. Then you get the more dedicated semi-professional people, and they'll go to the back of beyond; and they have to be pretty tough, and they walk hundreds of kilometers. That's a pretty limited bunch of people.

In Dargaville and elsewhere, says Beuse, collectors will scour a stretch of coastline for ambergris after every high tide, moving along the beach in three distinct waves. The first collectors are traveling at high speed, in cars—a practice that is legal on most New Zealand beaches despite a recent spate of accidents and deaths from collisions.

"They'd spot that going fifty kilometers an hour," she says, pointing to a small pale gray nugget of ambergris on the table. "It tends to sit proud on the high-tide line, so these people zoom along these high-tide lines looking for these pieces. It's about doing distance. You're going to wait a long time to get that large piece. In Dargaville, you're going to have to be moving like greased lightning at high-tide time, or you're going to be beaten."

The next wave moves more slowly, surveying the beach on motorbikes. "They'll chug-chug along after high tide and find smaller pieces," Beuse says. Finally, after the collectors in cars and on bikes have disturbed the sand, making off with the larger pieces, the walkers make their way onto the beach. Moving aside pieces of driftwood, they find the smaller pieces that have washed ashore and remain undetected. "It's a sense of thoroughness," she explains. "Dogs are used in the collection. It takes about a year to train a good ambergris dog."

Whether they rely on speed, churning up the beach after every high tide in a car, or use trained dogs and trek for days into remote and isolated backwaters, ambergris collectors take things seriously. "Here in New Zealand, it's almost a science," says Beuse. "They follow their currents and weather patterns. They're throwing a sandal in and following it and seeing where it ends up a week later. You know, practical experiments."

*   *   *

In 1825, in *The Handbook of Gastronomy*, the French gastronome Jean Anthelme Brillat-Savarin wrote:

It is well known that Marshal Richelieu, of glorious memory, habitually chewed ambergris pastilles; and as for me, whenever I feel, some day or other, the burden of age, when I think with difficulty, and feel oppressed by some power unknown, I take as much powdered amber as will lay on a shilling with a cup of chocolate, sugar it to my taste, and it has always done me a great deal of good. This tonic renders life more easy, makes thoughts flow with facility; and I do not suffer from that sleeplessness which is the infallible result of a cup of coffee taken with the intention of procuring the same effect.

On a cold drizzly day near the end of winter, finding myself in need of a tonic, I stand at my kitchen sink holding a teaspoon. In the bowl of the spoon is a little cairn of gray and black ambergris fragments from Taiwan—a small portion of a larger piece that was crushed in transit. It resembles cigarette ash now. After preparing a cup of hot chocolate, I pick up the spoon, drop it into the just-boiled liquid, and begin to stir. The ambergris disappears, replaced by a thick aromatic cloud of steam. I take two or three deep breaths. It is overpowering—floral, pungent, and grassy, with an aggressive sharp-edged underlying note to it.

Cautiously, I swallow a mouthful. A warm chewy lump of ambergris remains on my tongue. I bite down on it, and it sticks to my teeth. Unsuccessfully, I try to scrape it off, first with my tongue and then a fingertip. Remnants of it are still there two hours later—an oily black smear across one of my front teeth.

Around the rim of my mug, a thick black grainy ring of crushed ambergris remains. I sit calmly in my chair, in the gloom, waiting for some of the effects that Brillat-Savarin described. But I'm not sure they ever come.

*   *   *

Like traders of any other commodity, Beuse is a hostage to the vagaries of the market: to its soaring peaks and unanticipated troughs; the busy seasons, followed by the long slow months; the inescapable pressures of supply and demand. On occasion, she will receive so many requests to evaluate objects that people have found—which they hope are ambergris—that she is unable to answer them all. In September 2008, when the mysterious white lump washed ashore on Breaker Bay, Beuse says she watched the news reports with growing dread. For the next few days, her phone rang almost constantly.

"Sometimes," Beuse explains, "a big lot comes into the market and the prices are affected, but inevitably that supply will be exhausted. Then there's a real demand spike, because there's not enough around. And this is why the ambergris industry always lives in a state of tension."

Almost all of the ambergris she buys from collectors in New Zealand eventually will be sold overseas. "It's so expensive because it goes into a lot of different products," Beuse says. "There are a lot of end users. Beyond that, it's used in so much of the fragrance industry, throughout India and Asia. You've got religious associations with it in Arabic countries."

I suggest that the large French perfume houses—like Guerlain and Chanel—have replaced ambergris with synthetic ambergris compounds, such as Ambrox and Synambrane. "No," she says firmly. "No, no."

*   *   *

The largest piece of ambergris Beuse has traded, she says, weighed around twenty kilos—almost fifty pounds. I mime holding something the size of a large grapefruit in my hands. "Larger than that," she says, shaking her head. I slowly increase the distance between my hands until an imaginary basketball floats there. She shakes her head again. "It is important to know," she explains, "that New Zealand is not really known as a country that produces a quantity of ambergris, but New Zealand, because of its isolation, it's believed, and because perhaps of the diet of the whales that live around here: it's all pretty nicely cured here. We're about quality, not quantity. A big piece here in New Zealand is in the five- to ten-kilo range."

Beuse continues, "The higher the quality, the smaller the size. It reduces in size and rolls and knocks around and becomes better quality over time. If you come along with a hundred kilos, it's not going to be the best quality. New Zealand collectors generally get the best prices in the world. It's known throughout the industry. It's not liked, but it's known."

Sitting incongruously on the tabletop, among the ambergris, is a box of cold-and-flu medication. "I've got some beaks in here," Beuse says, picking up the box and tipping out a rolled-up piece of tissue paper, which she flattens to reveal a little black pile of squid beaks, like shiny blackened cinders. "There's a nice big beautiful one," she says, pointing out a long squid beak, curved like a claw and ending in a sharp point.

"This one is still sitting in the ambergris," she says, holding up a beak

Soft fresh ambergris. Credit: Christopher Kemp.

by its incurving tip so that I can see the rest of it, embedded in a little light-gray stratified cube of ambergris.

I tell Beuse I prefer the odor of the less-refined pieces. She describes them as "more fecal," but I find the aroma bold and sweet and enjoyable. Comparing its odor profile to the smaller white pieces of ambergris is like comparing the jolt of a strong Turkish coffee to the smoother, more subtle flavors of a café latte.

"I'll show you the really bad stuff, then," she says, pulling out another wad of tissue paper and slowly unwrapping a black shiny object. It is a little larger than a golf ball. "This is called soft ambergris, or fresh black ambergris, and this has been rolled into balls by a collector. They picked up one oozing mess." She hands it to me. "It's quite pliable," she says.

I squeeze it. It is soft, moist, and clammy, and it smells like fresh sheep droppings. The ambergris collector who found it and sold it to Beuse has left his or her fingerprints across its surface. It has been worked and shaped and kneaded like dough. I can see the whorls of a thumbprint on it.

Soft black ambergris is so fresh that it will melt if left in direct sunlight, says Beuse. "The demand for that product is considerably less," she

explains. "It's probably used quite a bit for burning for religious purposes. When collectors see that sort of soft black ambergris, it's associated
with whales being around in recent months or even weeks."

*   *   *

A day or so later, I attempt another dish, this time from Robert May's
1685 edition of *The Accomplisht Cook*. The recipe:

ANOTHER FORC'T FRYED DISH

Make a little past with yolks of eggs, flower, and boiling liquor.

Then take a quarter of a pound of marrow, half an ounce of cinamon, and
a little ginger. Then have some yolks of Eggs, and mash your marrow, and a
little Rose-water, musk or amber, and a few currans or none, and a little suet,
and make little pasties, fry them with clarified butter, and serve them with
scraped sugar, and juyce of orange.

I spend hours soaking and boiling beef bones, before removing the
gelatinous marrow with a teaspoon. Yellow and flaccid, like a softened
banana, it makes a wet sucking sound as it slides onto my plate. I mash it
with beaten eggs, cinnamon, and raisins into a pale fatty mixture, and use
a microplane to grate a small black piece of ambergris over it. The ambergris covers the surface like a layer of fine gray dust. I use the mixture to
fill little homemade ravioli, carefully sealing their edges with a fork before frying them in butter.

The end result is unpleasant—chewy and tasteless. The fatty unctuous
marrow filling has disappeared from inside the ravioli, rendered away
by the cooking process. The ambergris seems to have disappeared completely with it. There is no sign that it was ever there. The pastry reminds
me of cinnamon-flavored cardboard. I throw it away.

Instead, I scramble some eggs. Taking a small piece of white ambergris
that Beuse had given me, I grate it over a plate of steaming fluffy eggs. It
crumbles like a truffle. I fold it carefully into the eggs with a fork. Rising
and mingling with curls of steam from the eggs, the now familiar odor
of ambergris begins to fill and clog my throat, a thick and unmistakable
smell that I can taste. It inhabits the back of my throat and fill my sinuses.
It is aromatic—both woody and floral. The smell reminds me of leaf litter
on a forest floor and of the delicate, frilly undersides of mushrooms that
grow in damp and shaded places. Although present in only very small
amounts, the cholesterol-rich ambergris coats my tongue and the inside

of my mouth with a greasy film. I try to rinse it away by drinking water and eating dry slices of bread, but it remains for an hour or longer.

*   *   *

Back in the hotel room in Auckland, I'm holding a long thin piece of ambergris, which is almost white. It fills my hand. Its rough and pockmarked surface is covered with delicate white blooms and swirls of darker gray. I bring it to my nose and smell it. Even within the aromatic cloud that surrounds us, its odor is discernible immediately from the rest: it smells of the ocean. There is a thick briny freshness to it—the green, vegetal smell of wet seaweed. An hour earlier, when Beuse unwrapped it and placed it on the table, I was reminded of a museum curator handling an artifact.

I begin to understand it *is* an artifact.

It has taken decades to become the substance I am holding in my hand. In its complex odor is reflected every squall and every cold gray wave. I am smelling months of tidal movement and equatorial heat—the unseen molecular degradation of folded compounds slowly evolving and changing shape beneath its resinous surface. A year of rain. A decade spent swirling around a distant sinuous gyre. A dozen Antarctic circuits. In that moment, I finally understand why someone would be compelled to travel across the world to collect ambergris, and why it is has been so highly valued throughout history.

"Roughly, that piece should be about $1,000," says Beuse matter-of-factly. "I might be out a bit if there's more weight there. It's pretty much classic gray. It's high quality, not as refined as white. It's often very marine, very ocean, very fresh. You can smell the whales in that classic one there. You'll notice it's got an earthiness, but there's a marine quality there." She picks up a smaller white piece, cratered over its surface like a piece of a moon rock. "But if you smell this," she says, placing it beneath her nose, "you'll notice a sweetness. It almost has a vanilla to it."

I pick up another piece—a large dark pebble freckled with a starburst patina and mottled with dark brown and citrine-colored patches. It smells different again: grassy, like an old sundried cowpat. "That's pretty fresh," Beuse says, "and until recently it would have had black sticky material stuck to it."

The singularity of each piece of ambergris—its individual odor profile, the distinct color, and the mottling that covers its surface—serves as a reminder of the inadequacy of synthetic ambergris. "Every piece of this stuff smells different," Beuse confirms, "every single piece. If you have a one-kilo piece that breaks up on the beach, all of those pieces are likely to have the same properties, but if pieces break up in the ocean and float around, they are likely to have different experiences."

*　*　*

In 1811 Samuel Hahnemann published the *Materia Medica Pura: Volume 1*—an exhaustive 700-page compendium of naturally occurring drugs and their potential side effects. Hahnemann was the father of homeopathy. This was his bible. For each substance, Hahnemann painstakingly and obsessively collected hundreds of reports of side effects observed in patients and obtained from numerous different physicians.

There is a long section in the *Materia Medica Pura* devoted to ambergris and its effects. In total, physicians collected 490 separate observations following the administration of ambergris. Together they take up fifteen pages. The list is extensive. Some personal favorites:

Observation #76: Crepitation and creaking in the left ear, as when a watch is wound up; . . .

Observation #185: A frequent call to stool, but no motion occurs, and this makes her very anxious, and then the propinquity of other people is intolerable to her; . . .

Observation #452: Dreams full of business; . . .

Observation #487: His humour is easily embittered.

For several days after I eat a plateful of eggs and ambergris, when I first wake up in the morning, I lie in bed and take a quick inventory of my still-waking body. I ask myself: Is the propinquity of other people intolerable to me? Were my dreams full of business? Until now, I have noticed no effects. And then, this morning, when I awake in the half-darkness, I have an irresistible inclination to stretch (Observation #425).

*　*　*

Adrienne Beuse methodically gathers her ambergris from the tabletop, carefully wrapping each piece again in cotton. She slowly places them back into her bag. At the door to my room, we shake hands, and she turns, trudging away toward the elevator. In same moment that I close the door behind her, another door—which had opened momentarily on a strange and secretive world populated by ambergris traders like Beuse and Bernard Perrin, and hunters like John Vodanovich—closes too.

After Beuse is gone, the thick musty smell of ambergris still hangs in the air. I try to imagine her riding the elevator to the lobby and pushing through the Italian tourists, her tote bag clutched tightly beneath her arm.

I walk to the window and part the curtains. Late afternoon. It is still bright and warm outside. To the north, the water glitters in the distance. A long line of cars snakes across the bridge, frozen in light, glinting in the sun. I look down toward the sidewalk four stories below, waiting for Beuse to exit the hotel entrance and join the crowds on the streets. Hoping that she might somehow give away more in an unguarded moment than she had in the hours we had just spent together, I decide to watch her for a moment—crossing the street, navigating the city, her tote bag filled with ambergris—unaware she is being watched.

Fixing my eyes on the empty space between two potted plants below, I stand at the window. And I wait. The water shimmers in the blue bowl of the harbor. The cars speed toward downtown Auckland. But Beuse never appears.

EPILOGUE

It's a rainy morning at Aramoana. The air is wet against my face. I watch the water as it exits the harbor, sliding past the green hills. The tide is oceangoing. I'm standing on the spit with my son. He is eighteen months old now and walking. Across the water, on the grassy cliff tops at Taieri Head, the lighthouse stands like a red-topped tabernacle.

From somewhere—another beach perhaps, or even another continent—the last tide has brought to shore a multitude of tiny white flowers. They are everywhere. Wet clusters of them have collected like confetti in the green folds of the kelp. A thin white seam of them demarcates the high-tide line, stretching irregularly along the beach as far as I can see. We follow it until we grow tired. To our right, the cliffs rise like a weedy overgrown wall.

As it always does, the sea has carried waterlogged pieces of driftwood and dumped them on the shore with a multicolored collection of gloves, oyster shells, car tires, fish bones, broken buoys, and crab claws. I pick up a blackened stubby ear of corn. Most of its kernels have been removed by the sea—it reminds me of a mouth, missing teeth. A little farther along the beach, I drop the corn on the sand and exchange it for a red plastic water pistol.

Removing a bag from my pocket, I collect a handful of the elegant, thin spire-shaped shells that gather here after high tide. We will take them home. This morning, they are reason enough for being here. Once again, I have not found ambergris. But, all these months later, I no longer expect to find any. In fact, my search for ambergris is almost over. A few weeks from now, after more than two years in Dunedin, we will leave New Zealand and return to the United States.

*   *   *

In other subtler ways, my journey had begun ending months earlier. On my desk at home, a clutter of accumulated marginalia: a large, pitted lump of pumice; a dried sea horse; a vial of perfume that contains ambergris, and another that was made with synthetic ambergris; a two-foot-tall stack of seventeenth- and eighteenth-century monographs on ambergris; large scale maps of Stewart Island and the state of Massachusetts; an old dog-eared paperback edition of *Moby-Dick*; newspaper clippings; rock samples; and several different and varied samples of ambergris from across the world.

I began this journey because I had so many questions, and it seemed there were so few answers. Finally, I felt that I knew everything there really was to know about ambergris. I had exchanged ignorance for knowledge. I am, after all, a scientist. This is what I was trained to do.

Along the way, I had seen numerous pieces of ambergris. I had handled and smelled it firsthand—in museums in New Zealand, on Stewart Island where Department of Conservation workers were finding rounded pebbles of it on remote west coast beaches, and then farther afield. I'd met with ambergris vendors like Adrienne Beuse and cultivated relationships with full-time ambergris collectors like John Vodanovich. As the months passed, I spoke with museum curators, perfumers, oceanographers, cetologists, and organic chemists. I had even eaten ambergris, attempting to re-create elaborate recipes from seventeenth-century England.

In other words, finding ambergris no longer seemed like an appropriate measure of success. At some point—maybe on Stewart Island; or along the wintry shoreline of Cape Cod; or perhaps somewhere in between—the journey had become the destination instead.

*   *   *

My search for ambergris has allowed me to know the natural world more fully than I otherwise would have. Along the wild Otago coastline, the sky is most beautiful just before a storm. Slate-gray thunderheads begin to form, drifting in from the sea and pressing against the green cliffs. They pile on top of one another—enormous black columns that tower thousands of feet into the sky. And then the rains come, sweeping across the beach in wet gusts, dimpling the sand with quarter-size craters. I might never have known this if I hadn't been searching, week after week, for ambergris. Before the mysterious object washed ashore on Breaker

Bay in September 2008, I had not been accustomed to standing on remote coastlines, as violent storms develop out to sea. On rainy mornings,
my son and I would sit in front of our living-room window, watching as
fat raindrops wobbled down the telephone lines that crisscross our hilly
street. Now we reach for our raincoats instead.

My journey has been a circular one. It had begun here, on the wet sand
at Aramoana under a pearlescent sky, and farther north at Long Beach. It
seems fitting that it should end here too.

Low tide. A mile or so south of here, a wide sward of green land is
emerging from the middle of the harbor, as it always does when the tide
recedes. Its rocky clay-colored edges appear first, like the crenellations
atop a crumbling keep—the walls of a lost city. And then comes a broad
table of flat land stretching out toward the sea, pristine and crowded with
birds. I often imagine rowing across the harbor, waves slapping at the
bow, until I arrive at the exposed green apron and step gingerly from the
rocking boat, onto the leafy seaweed—feeling it spring softly beneath my
feet like wet lettuce. I have discussed this several times with my wife—my
eternally patient wife—who is always willing to thoughtfully debate the
most ridiculous questions. Will the wet new land in the harbor bear my
weight? Yes, she believes it will. And might I find ambergris out there in
the quiet air, after all the water is gone?

No one knows. But, for an ambergris collector, the only constant is
hopefulness in the face of incredibly low odds.

*   *   *

Back on the beach at Aramoana, the birds are on the water. They ride the
rolling swell. I walk past wet green concertinas of wakame with my son,
prying up large pieces of driftwood to look beneath them. The storm is
approaching, faster now. Gusts of wind flatten the yellow dune grass. I
carefully pocket the shells still in my hand—which I now know are a
member of the Turritella genus—and we begin the slow trek across the
sand back to my car. It's time to go home. I take my son's cold hand in
mine. We walk past a quick-legged pair of oystercatchers patrolling the
margins, and I see a pale rock on the beach that looks different from the
rest. I'm drawn to it. It sits next to a toothbrush and a tangle of kelp, half-
buried in the sand, shaped almost like a duckbill, white and chalky at one
end, mottled with dark striations at the other. I bend to pick it up, still
wet, and bring it cautiously to my nose.

And I smell it.